to
Eileen
Emma and Michael

Never measure the height of a mountain until you
have reached the top. Then you will see how low
it was.

From the diary of Dag Hammarskjold (1905 - 1961), Swedish diplomat and
political economist.

Understanding Mathematics

Keith Gregson

NOTTINGHAM
University Press

Nottingham University Press
Manor Farm, Main Street, Thrumpton
Nottingham NG11 0AX, United Kingdom
www.nup.com

NOTTINGHAM

First published 2007
©K Gregson

British Library Cataloguing in Publication Data
Understanding Mathematics
K Gregson

ISBN-10: 1-904761-54-2 ISBN-13: 978-1-904761-54-9

Disclaimer

Typeset using LaTeX.
Printed and bound by Hobbs the Printers, Hampshire, England

Contents

7 Integral Calculus

Preface

This book is written for those of you who are struggling with mathematics, either as first year undergraduates taking maths as a subsidiary to a science course, or students working for S- and A-levels. It is not a comprehensive A-level syllabus text - there are many such books and software courses.

There are some students who take to mathematics easily, there are others (the majority) who have difficulty. This difficulty is generally the result of having a hard time throughout earlier education. I know that many of you have been told repeatedly that mathematics is hard. Much of it is not - but mastering it does involve a different approach, memorising a few formulae is not the answer. Success comes from reading, several times if necessary, together with thinking, until the mathematics is understood. Sometimes understanding doesn't come easily. You will need to be able to seek help - ask your fellow students, or tutors, or whoever. This takes effort - **from you**, but what you put in will pay great dividends, so don't give up without a fight! Attitude of mind is important and succinctly put by Henry Ford - "Whether you believe you can do a thing or not, you are right."

The book is made up from notes which I have given to first year undergraduate students in biosciences, though it could be equally useful to students in other areas of science and engineering. Its aim is to encourage the student to appreciate how mathematical tools are derived from a few simple assumptions and definitions, and thereby to discourage the examination habit of memorise, recall and forget. The emphasis is to encourage students to think about a few topics in order to help them develop an understanding of the underlying principles of mathematics.

Read the book steadily, a little at a time, working through the proofs and examples until you are confident that you understand them. Try the exercises, but don't spend more than twenty minutes on any of them. If you can't solve them, try to identify the difficulty so that you can ask for help, then ask!

Most mathematical formulæ can be found from first principles without too much difficulty and the act of derivation promotes understanding and confidence. That is why they are included, not so that you can remember them. If you have followed the argument once, that is sufficient. If you use the formula subsequently you will be surprised what you can remember.

Difficult formulæ are best looked up, but it is helpful to have sufficient understanding to know where to start looking. It is also important in this Internet age to know that what you are looking at has some authority.

I have deliberately avoided the rigorously inflexible approach which is essential to a pure maths readership. Rather, I have attempted to develop the spirit of the subject without examining all the details. As in life, it is often better not to question every assumption, lest it detracts from the more important (and fun) moments. As new topics are introduced I have gradually reduced the verbal description, in line with the (I hope) growing familiarity of the reader to handle the notation and brevity.

The idea for the book came from some of my students and I hope that the result will both please them and be useful to others. I thank them for their belief.

I must also thank my wife for helping me find the time (by taking on all the other things that I should have been doing!), and my family simply for being themselves - I count myself lucky.

As a final note I would like to express my gratitude to the staff at Nottingham University Press; in particular to Sarah Mellor for her enthusiasm and encouragement.

Keith Gregson
February 2007

Chapter 1

Fundamentals

1.1   Why Mathematics

It is interesting to see the approach that newspapers, even the so called informed and respectable press, have toward the current craze for Sudoku. Almost without exception they will tell you that Sudoku problems can be solved without recourse to mathematics. They qualify this statement with the further claim that the puzzles may be solved merely with reasoning and logic. They clearly do not understand what mathematics is about - so if you are having difficulty, you are not alone!

It is true that Sudoku problems are not numerical, you could just as easily substitute a,b,c...for 1,2,3...- but mathematics is much more than arithmetic - logic and reasoning for example?

They are all perpetuating the myth that mathematics is hard and should be avoided at all costs. That some mathematics is difficult is undeniable, but much of it is not. So where does the myth come from?

Some of it arises because we (most of us) do not like to be fenced in by rules and regulations - and maths depends upon rules, though there are surprisingly few of them. You can't simply scribble a few lines of "mathsy" stuff on paper and expect the readers to shower praise on you, your stuff has to stand up to criticism and analysis - it's either right, or wrong.

The myth is also partly the result of the current education system. Mathematics syllabuses are too full and therefore rely far too heavily on memory. Students know a great deal, but they are given too little time to explore and develop **understanding**. If you know how it works, it is surprising how often you can apply the same principle in different situations, and when you do, it is very satisfying to know that **you** solved the problem.

1.2   What's it all about?

1.2.1 x?

Next to your signature it's affectionate; it registers a vote; eight of them might win you a fortune on the football pools. It's one of the easiest letters to write, hence it was used as a signature; I never did understand that - if anyone could write it, how did you know who did?

But put it in an equation and it causes nausea and sleeping sickness. This is a shame, because that's exactly what it's not supposed to do. So! what the L is an x then? Well, first of all you will notice that we write x like this - x, not like this - x - because this can easily be mistaken for a multiplication sign. It's a symbol which stands for "an unknown quantity of something". We sometimes use y to represent "an unknown quantity of different somethings", so you see we can, and do, use other symbols.

In English we use complicated symbols (words) to represent things: for example we use 'fire-engines' to represent a group of mobile things which are used to put fires out. You see the symbol 'fire-engine' and you picture a red thing on wheels with a ladder and lots of big hose-pipes. But life isn't quite so simple: if you lived in the USA you would think of similar things, only yellow and if you lived somewhere else you might think of a wheelbarrow with a bucket in it! So you see, even symbols which we think we understand are never quite so clear cut.

So why do we use x? Well isn't it easier to write

$$x^2 + 3x + 5 = 0$$

rather than

$$\text{fire-engine}^2 + 3 \times \text{fire-engine} + 5 = 0$$

or

$$something^2 + 3 \times something + 5 = 0?$$

We use x a lot because it's easy to write, and it saves us a lot of time and effort, <u>that's all</u>.

1.2.2   Mathematics?

Mathematics is a language, like English or Spanish or Russian. It allows us to express relationships, not simply brother and sister and parent type relationships, but how things depend on other things. For instance, the area of a rectangular garden can be written

$$area = length \times width$$

except of course, being mathematicians we'd write

$$a = l \times w$$

because we prefer to write as concisely as possible. In fact we would probably write $a = lw$ because that's even more economical - we often leave out the multiplication sign when the context is obvious.

Mathematics is also a discipline, it makes us justify everything that we do: what are the assumptions? what can we derive from the assumptions? and can we prove the results? And perhaps more importantly, especially if you are an applied mathematician - which we all are - how can we interpret and use the results?

1.2.3 Functions and Equations

A *function* is a relationship between two or more variables. One of these variables (the *dependent* variable) is defined in terms of the other variable/s (the *independent* variable/s). When we know what the relationship is we can normally write it down as an equation. So, for example, the area of a circle may be represented by the equation

$$A = \pi r^2$$

in which A, the dependent variable, is defined in terms of r, the independent variable.

When we need to calculate the value of the dependent variable we substitute values for the independent variables and evaluate the resulting equation. This sounds complicated because we are trying to behave like mathematicians. However, consideration of a simple example should make things easier. We can calculate the area of a circle of radius 10cm by substituting the value of π and the value for r as follows:

$$\begin{aligned} A &= 3.14159 \times (10cm)^2 \\ &= 314.159cm^2 \end{aligned}$$

Note that we often omit the units from equations, because they can confuse the issue. However we should always remember to include them in the final answer, and we can use them (or more particularly their dimensions) to advantage when checking equations. (see dimension analysis, page 9).

Now you could say "So what?" or "What's the big deal" and I would reply:

> "Well, I don't need to remember that the area of a circle with a radius of 10 is 314.2, or that the area of a circle of radius 2 is 12.6. Neither do I need a table of such values, nor a carefully plotted calibration curve. All that is necessary is the equation $A = \pi r^2$."

Think of the time, effort and paper that this function saves! Now think of all the other formulæ that you have used and imagine life without them.

One big advantage of mathematics is that you can write something like $A = 3.14159r^2$ and be understood anywhere in the world. Try writing this relationship in English! Indeed, if we were to send a message into space, in the hope that some *alien intelligence* will find it and realise that there is intelligence elsewhere, we would stand a much better chance of success if we transmitted the message "$A = 3.14159r^2$" rather than "roses are red". Think about it! (the number π has a value which we believe is a universal constant) And while you are in thinking mode - have you noticed that the formula $A = \pi r^2$ is very similar to Einstein's $E = mc^2$?

1.2.4   Relationships

Hopefully you're beginning to get the idea. We use mathematics to express how things relate to each other. As scientists we spend our time looking for these relationships (e.g. $A = \pi r^2$). Isn't it therefore sensible to be able to write them down clearly, so that others can share our discoveries and use them to extend the general knowledge in order to build things (like iphones) and solve new problems? If you have done the hard work you may as well document it in the best possible way.

1.2.5   Why don't we speak mathematics all the time?

Well, whilst it's good for describing precise scientific things, it's not too good for chatting about everyday things like football and dancing and pop-music and stuff - for that we prefer a more evocative language in which we can exercise a bit of imagination.

English can often be interpreted in different ways. For example, the phrase "shed load", which occurs in the context of road traffic reports makes us smile - "How big a shed?". And a report on the problems of increasing weight brought forth the phrase "$\cdots$ the ballooning weight problem $\cdots$" to which a reporter interjected "What is the optimum weight for ballooning?".

Mathematics isn't like that, when we speak (or more usually write) maths, we are trying to be precise and unambiguous. We want to be clearly understood and concise.

1.2.6   .. and why do I need to understand it?

A couple of simple examples might just help.

Suppose you go to the shop and buy a packet of soap at £3 and two bottles of shampoo at £2 each. You watch the cashier type in $3 + 2 \times 2$ and the till shows that you owe £10. Do you pay it? First of all try typing the expression as it stands into your calculator. Depending upon the age and type of your calculator you will produce one of two answers; either £7 or £10. Which is correct? If you do, your mathematics is OK. If you don't, you are likely paying a lot more than you need! In any event don't you think you owe it to yourself to check these things now and again? (If you don't please could you send the author a cutting from the tree on which your money grows!)

As another example: suppose that your financial advisor suggests that you should take out a bond which will give a 60% return after ten years rather than invest at 5% per year for ten years, after all - 10×5 is only 50! What should you do? If you know the answer to this one your mathematics is very good. But do you understand the calculation? If you do you may be able to skip a few chapters, otherwise read on.

1.3 Working with Equations

An equation consists of two expressions separated by an equality sign $(=)$, though sometimes the separator may be an inequality sign (e.g. $<=$).

1.3.1 Rearranging Equations

There are lots of so called rules regarding the manipulation of equations: "Ignore them!" All that you need to remember is that you can do almost anything you like to an equation - provided that you **treat both sides of the equation in the same way**. So that you can add 10 to **both** sides, subtract 23 from **both** sides or add x to **both** sides. Similarly you can multiply or divide **both** sides by anything, take the logarithm of **both** sides or whatever. Just be careful about multiplying or dividing by zero, that's all.

So, for example, if we start out with the equation

$$y = mx + c$$

and would like to find an equation for x we could subtract c from both sides to give

$$y - c = mx + c - c$$
$$y - c = mx$$

and then divide both sides by m to give

$$\frac{y - c}{m} = \frac{mx}{m}$$

then, by cancelling on the RHS (right hand side)

$$\frac{y - c}{m} = \frac{\cancel{m}\,x}{\cancel{m}}$$

and writing the equation in reverse order we get

$$x = \frac{y - c}{m}$$

The **only** thing to remember is: whatever you do to one side of the equation, you must also do to the other. **But** beware of multiplying or dividing by zero, because; while $0 \times 2 = 0 \times 3$ is true, the result of dividing both sides by zero $(2 = 3)$ **is not!**

The use of this simple trick allows charlatans to accomplish all manner of things!

1.3.2   Order of Evaluating Algebraic Equations

When we evaluate numerical expressions such as $3 + (2 + 3)^2$ we do not, in general, work from left to right, see the following examples. Mathematical operations are carried out in a strict order as defined by the priority list in *table* (1.1). Operators high in the list are completed before those lower down. When two operators have the same priority it is customary (but not mandatory) to work from left to right. (In computer programs this may not be the case.)

priority	operation
1	brackets ( )
2	indices, powers, exponents
3	$\times, \div, /$
4	$+, -$

Table 1.1: Priority of Mathematical Operators

Be especially careful with expressions like $a \div b \div c$ since they can be easily misinterpreted. Does it mean $a \div (bc)$ which results from evaluating $a \div b$ first or $(ac) \div b$ when $b \div c$ is evaluated first? In cases like this you should use brackets to make things clear.

But don't develop the habit of using brackets unnecessarily either, a plethora of brackets can be more confusing than none at all.

Example - Evaluate $3 \times (2 + 3)^2$

$$
\begin{aligned}
3 \times (2 + 3)^2 &= 3 \times (5)^2 \\
&= 3 \times 25 \\
&= 75
\end{aligned}
$$

Example - Evaluate $3 \times 2 + 3^2$

$$
\begin{aligned}
3 \times 2 + 3^2 &= 3 \times 2 + 9 \\
&= 6 + 9 \\
&= 15
\end{aligned}
$$

1.3.3 Some Useful Algebraic Relationships

$x + y = y + x$	commutative law of addition
$x + (y + z) = (x + y) + z$	associative law of addition
$x \times y = y \times x$	commutative law of multiplication
$x \times (y \times z) = (x \times y) \times z$	associative law of multiplication
$x(y + z) = xy + xz$	
$x + 0 = x$	definition of zero
$x \times 1 = x$	definition of unity
$\frac{x+y}{z} = \frac{x}{z} + \frac{y}{z}$	
$\frac{x}{y} \times \frac{v}{w} = \frac{xv}{yw}$	
$\frac{x}{y} + \frac{v}{w} = \frac{xw+yv}{yw}$	

Table 1.2: Algebraic Relationships

1.3.4   A word about Calculators

Calculators, like computers, can save a lot of work. Indeed they allow us to perform tasks which are impossible with only pen and paper. However, as with all things, there are disadvantages. The main of these is that they take us (the users) a little bit further from understanding the processes of arithmetic. Here are a couple of problems which I have come across when talking mathematics with students.

1. Most modern calculators understand the priorities that govern the order in which calculations are carried out. Older calculators didn't! so if you've inherited your calculator from Mum or Dad, be careful! In the old days **you** were expected to enter the problem in the correct order. If in doubt about your calculator, try entering the following in the order presented:
$$2 + 3 \times 5 =$$

   On a modern machine you will get 17 (correctly). On an old-type calculator you would produce an incorrect result of 25, the reason being that the calculator performs the operations as you enter them:
$$(2 + 3 \Longrightarrow 5) \times 5 \Longrightarrow 25$$

   In order to obtain the correct result on such a machine you should type the problem as
$$3 \times 5 + 2 =$$

2. Misunderstanding of the "EXP" key frequently results in errors.

   2 EXP 3 means 2×10^3 **not** 2^3 and will probably appear on your calculator as 2E3

   2^3 (i.e. $2 \times 2 \times 2$) is usually performed using the $\wedge$ or x^y key as follows:
$$2 \wedge 3 = 8 \qquad \text{or} \qquad 2 \; x^y \; 3 = 8$$

If in doubt try a simple calculation for which you know the answer.

1.4   Preliminary Calculations - check the problem

It is very tempting to rush in to a computation with calculator in hand. However, it is often useful (and safer) to perform some simple arithmetic on the back of an envelope first.

There have been many cases where a quick check of the calculations would have avoided serious consequences - it is very easy to make mistakes when keying in numbers, or to misinterpret results from a calculator or spreadsheet. When such calculations may have serious consequences, you should check. Miscalculating a drug dose by a factor of ten is a serious (and real) example.

There is also, of course, the possibility that your computer or calculator may produce the wrong answer! How dare anyone even think such a thing?!

I shall illustrate what can be done by looking at an example from geochemistry. Geochemists use a relationship called *residence time* in order to elicit information about the movement of chemical elements through large systems like the oceans.

$$\text{residence time} = \frac{\text{mass of element in oceans}}{\text{mass turnover per year}} \qquad (1.1)$$

$$= \frac{\text{conc'n in oceans} \times \text{volume of oceans}}{\text{conc'n in rivers} \times \text{flux from rivers}} \qquad (1.2)$$

Residence time is the length of time that an element remains in the ocean.

1.4.1 Dimension Analysis

A simple check which we can perform on formulæ like *equation* (1.1) above is to examine the dimensions involved. Note here that we are talking about **dimensions** (e.g. mass, length, time, electric current, ...), <u>not</u> **units**. In order to carry out this check we replace the terms in the formula by their dimensions (mass(M), length(L), and time(T)).

$$\text{residence time}(T) = \frac{\text{mass of element in oceans}(M)}{\text{mass turnover per year}(MT^{-1})}$$

In this case, after cancelling the M's on the RHS, the dimensions on both sides of the equation agree (T), so that we can be reasonably confident that the formula will generate a sensible answer. When the dimensions do not agree the formula is in error and should be checked.

1.4.2 An example calculation

One question that a geochemist might like to ask is: "What is the annual flux of rainwater into the oceans?" This could be achieved using the quantities and concentrations of calcium (Ca) as follows:

Residence time of Ca in the oceans	0.818×10^6 years
Concentration of Ca in seawater	412 mg dm^{-3}
Concentration of Ca in river water	15 mg dm^{-3}
Volume of the oceans	1.37×10^{21} dm^3

First we must rearrange *equation* (1.2) as follows:

$$\text{flux from rivers} \;=\; \frac{\text{conc'n of Ca in oceans} \times \text{volume of oceans}}{\text{conc'n of Ca in rivers} \times \text{residence time}}$$

$$=\; \frac{412 \times 1.37 \times 10^{21}}{15 \times 0.818 \times 10^{6}}$$

1.4.3   And the back of the envelope

Evaluating the above expression will have most of us diving for a calculator or spreadsheet. However it is a good idea to perform a rough evaluation on a scrap of paper first, since it is easy to make a mistake entering the values. We do this by approximating the individual values to the extent that we can perform a rough calculation using head and pencil only.

You will just have to imagine that the following is a scruffy envelope!

$$\approx\; \frac{(4 \times 10^{2}) \times 1.4 \times 10^{21}}{15 \times 0.8 \times 10^{6}}$$

$$=\; \frac{4 \times 1.4}{15 \times 0.8} \times \frac{10^{2} \times 10^{21}}{10^{6}}$$

$$=\; \frac{5.6}{12} \times 10^{17}$$

$$\approx\; \frac{1}{2} \times 10^{17} \text{dm}^{3}\text{year}^{-1}$$

Performing the calculation correctly on a calculator produces the result 4.6×10^{16} (or 0.46×10^{17}). The agreement between this and the rough answer should make you feel confident that it is correct.

Exercises

1. If
$$\frac{a}{b} = \frac{c}{d} + e$$
   find expressions for each of the five variables in terms of the other four.

2. The equations of motion for a body acted upon by a constant force are:

$$\begin{array}{rcll} v & = & u + at & (1.3) \\ v^2 & = & u^2 + 2as & (1.4) \\ s & = & ut + \frac{1}{2}at^2 & (1.5) \\ s & = & \left(\frac{u+v}{2}\right)t & (1.6) \end{array}$$

   where u is the initial velocity, v the final velocity, s is the distance travelled, t is the time, and a the force applied.

   (a) find an expression for a from equation 1.3.

   (b) find an expression for u from equation 1.4.

   (c) find an expression for a from equation 1.5.

   (d) find an expression for v from equation 1.6.

3. The time in seconds taken by a pendulumn to swing back and forth and return to its original position is known as its period and it may be calculated from the following expression.

$$t = 2\pi\sqrt{\frac{l}{g}}$$

   where t is the period (s), l is the length of the pendulumn (m) and g is the gravitational constant (9.81 m s^{-2}).

   Find an expression for l in terms of t and g, and hence calculate the length of a pendulumn that will have a period of one second.

   How might we use the formula to calculate the gravitational constant on the moon, should we be fortunate enough to get there?

Answers

1. (a) $a = b(\frac{c}{d} + e)$

   (b) $b = \frac{ad}{c+de}$

   (c) $c = (\frac{a}{b} - e)d$

   (d) $d = \frac{bc}{a-be}$

   (e) $e = \frac{a}{b} - \frac{c}{d}$

2. (a) $a = \left(\frac{v-u}{t}\right)$

   (b) $u = \sqrt{v^2 - 2as}$

   (c) $a = 2\left(\frac{s-ut}{t^2}\right)$

   (d) $v = \left(\frac{2s}{t}\right) - u$

3.

$$l = \left(\frac{t}{2\pi}\right)^2 g$$

For a period of one second the length of the pendulumn should be 25cm.

Measure the length and period of a pendulumn and use the formula

$$g = l\left(\frac{2\pi}{t}\right)^2$$

Chapter 2

Numbers

We are all familiar with the natural numbers $1, 2, 3 \ldots$ Doubtless these numbers came about because of the need to count when bartering or trading. They are referred to as *natural numbers*, or *positive integers* or in every day parlance as "whole numbers".

Throughout history our use of numbers has developed in many ways and many different number systems and notations have been devised. The evolution of numbers is like the curate's egg: *did numbers exist before the need to count, or were they invented because of a need?* Whatever the answer to this question you should be aware of some of the different number systems and representations, and of their uses.

We shall concentrate on the decimal number system, since this is the most common and convenient system for humans with a full compliment of fingers!

Besides the positive integers, there is another set comprised of the *negative integers* (corresponding to a debt?) which, together with the number "zero" are collectively called *integers*.

The integers suffice when we are in counting mode, but have severe limitations when measuring, or if we need to perform anything but the simplest of calculations. Thus we need another set of numbers which are the *real numbers*. These numbers are the set of numbers familiar to scientists, since they are the results of our calculations and measurements. They in turn are made up of two sets of numbers; *rational numbers* which, as their name implies, are made up of ratios or fractions like $0.5, 0.3\dot{3}$ and $2/3$, and *irrational numbers* which are a special set of numbers which can not be expressed as ratios - this set includes numbers like π, e and $\sqrt{2}$; they are all infinite unrepeating decimal fractions. Note that the set of *real numbers* includes all the *integers*.

The set of real numbers allows us to solve most of the mathematical problems with which we shall be faced. It should be remembered however that there are problems which may only be solved by recourse to a wider set of numbers - e.g. many electro-magnetic problems can only be solved in terms of *complex numbers*. (complex numbers are imaginary things which help us to think about the square roots of negative numbers, that's why they are often referred to as *imaginary numbers*)

Real numbers may be represented by points on a straight line:-

This line is referred to as the *real number line*. The dashes at each end indicate that it may be extended in either direction indefinitely.

Drawing infinitely long lines or writing indefinitely large numbers is impossible! However it is sometimes necessary to represent these concepts and so we have invented the symbol ∞ which represents an unimaginably large (and hence uncountable) number which we call "infinity". We could redraw the real number line using this symbol as follows:

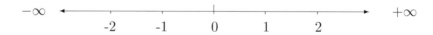

We must be careful not to use the symbols $\pm\infty$ to refer to specific values, they are simply symbols of hugeness beyond measure!

2.1   Decimal Number Representation

Some examples of numbers (in decimal notation) are:-

$$234$$
$$3.14159$$
$$2.\dot{6}$$
$$5000000$$

The over-scored · in the third number indicates that the 6 should be repeated indefinitely, an alternative way of writing this would be $2.666\cdots$. An even better and more concise form is $2\frac{2}{3}$. It is interesting to note that this number can never be expressed accurately in decimal; some numbers are like that - indeed you should be aware that the vast majority of numbers can not be represented accurately in a computer (why not? [1]).

We live in an approximate world - a place that mathematicians and physicists refused to believe in for a long time!

---

[1]because numbers in a computer are stored using a fixed number of significant figures - usually about 7. Therefore the best that could be done with $2\frac{2}{3}$ would be 2.666667.

2.1.1 Significant Figures and Decimal Places

The second of the numbers above represents π to 6 significant figures, and to 5 decimal places, because it contains 6 digits in total and has been rounded to 5 digits after the decimal point. The last number has been represented to 7 significant figures and <u>no</u> decimal places. The third number is represented to an infinite number of significant figures and decimal places - but we don't normally do that sort of thing, and we certainly cannot on a computer [1].

2.1.2 Scientific Notation

Accountants would write the last number as 5,000,000 in order to assist the reading. We (scientists) don't! - though we often deal with very large or very small values. The way we get around the problem of writing such numbers is to use "scientific notation" in which we would represent the number five million as:

$$5.0 \times 10^6$$

or sometimes, especially if the number is on a calculator or a computer screen as $0.5E7$ or $0.5 + 7$ or $0.5_{10}7$. There are several variations on these, depending upon the calculator or computer program in use, but basically the number is represented either by a fraction or a number between 1 and 10 (the *mantissa*, 5.0 in this case) multiplied by a power of ten (the *exponent*, 6 in this case). Remember that multiplying by 10 shifts the decimal place one digit to the right, multiplying by 10^2 shifts the decimal place two digits to the right and so on. If the power of ten is negative we shift the decimal point the opposite way *ie* to the left. Here are some numbers expressed in this way:

distance of the Earth from the Sun	1.5×10^{11} m
	(150000000000 m)
speed of the Earth around the Sun	2.98×10^4 m s^{-1}
	(29800m s^{-1})
diameter of the Earth	1.2756×10^7 m
	(12756000 m)
mass of the hydrogen atom	1.67×10^{-24} g
	(0.00000000000000000000000167 g)
speed of light	2.9979×10^8 m s^{-1}
	(299790000m s^{-1})
Avogadro's number	6.02252×10^{23}
	(602252000000000000000000)
Age of the Earth	4×10^9 years
	(4000000000 years)
Age of the Universe	1.5×10^{10} years
	(15000000000 years)

2.2 Binary and Hexadecimal numbers

Binary and hexadecimal number representation is fundamental to understanding how computers store data. All modern computers use binary notation in order to represent and operate upon numbers. A brief understanding of binary will therefore be helpful in understanding how computers work, while hexadecimal provides a convenient representation of large binary numbers. You may have no interest in the internal workings of a computer, and I can sympathise with that, in which case you could safely skip this section, though you may still find it enlightening to see how numbers can be represented using a different base.

The base or radix of a number system is the number of different digits, including zero, that the system uses. The decimal system uses ten different digits and all numbers are represented as a sequence of powers of ten. As a reminder consider the number:

$$538$$

which we interpret to mean:-

$$
\begin{array}{ll}
5 \times 10 \text{ to the power } 2 \ (\ 5 \times 100 = 500 \ ) \\
+ \ 3 \times 10 \text{ to the power } 1 \ (\ 3 \times 10 \ = \ 30 \ ) \\
+ \ 8 \times 10 \text{ to the power } 0 \ (\ 8 \times 1 \ \ = \ \ 8 \ )
\end{array}
$$

This system has been accepted more by accident than for logical reasons. In fact there are many arguments in favour of different number systems but until the advent of the computer there was little pressure to understand any of the alternatives. Why then, should we consider them now?

Fundamentally the major difficulty in operating with decimal numbers is the necessity of providing ten symbols to allow the representation of numbers. These symbols are of course the familiar digits 0,1...9. If we were to build a computer based on this system we would need electronic devices which could distinguish between ten different states. This is difficult! It is much easier (and more reliable) to build electronic devices which recognise two states e.g. current is either flowing or not, a magnetic field is either polarised in one direction or the other, or a switch is on or off. For this reason computers are built using binary logic in which all information is stored as strings of 1's and 0's, and so if we are to understand the working of a computer, it will be helpful to have some knowledge of binary arithmetic.

2.2.1 Binary Numbers

We are used to representing large numbers as a sequence of powers of 10. For example 231 means $2 \times 10^2 + 2 \times 10^1 + 1 \times 10^0$. In binary we represent numbers as a sequence of decreasing powers of 2 so that the binary number is interpreted as

$$
\begin{array}{rl}
 & 1 \times 2 \text{ to the power } 4 \\
+ & 0 \times 2 \text{ to the power } 3 \\
+ & 0 \times 2 \text{ to the power } 2 \\
+ & 1 \times 2 \text{ to the power } 1 \\
+ & 1 \times 2 \text{ to the power } 0
\end{array}
$$

and is the equivalent of the decimal number 19.

It will be obvious that, though we need only two symbols (0 and 1) in binary, in general the binary representation of a number will require more digits than the equivalent decimal representation. A few examples may be helpful.

decimal	binary
5	101
0	0
65	1000001
31	11111
10.5	1010.1
3.25	11.01

Within binary numbers the "." is known as the *binary point*.

2.2.2   Conversion from Decimal to Binary

In order to convert the decimal number 13 to binary we repeatedly divide by 2 as follows:-

$$
\begin{array}{ll}
13 \; / \; 2 = 6 & \text{remainder } 1 \\
6 \; / \; 2 = 3 & \text{remainder } 0 \\
3 \; / \; 2 = 1 & \text{remainder } 1 \\
1 \; / \; 2 = 0 & \text{remainder } 1
\end{array}
$$

Now if we read the remainder column starting from the bottom we have 1101 which is the binary equivalent of the decimal number 13.

2.2.3   Hexadecimal Numbers

In the hexadecimal system we use the number 16 as the base (or *radix*) as opposed to 10 in the normal decimal system or 2 in binary. This means that we have to 'invent' 6 new symbols to represent the additional digits. The hexadecimal digits are as follows:-

$$
0, 1, 2 \ldots 9, A, B, C, D, E, F
$$

so that

$$A_{16} = 10_{10}$$
$$B_{16} = 11_{10}$$
$$\vdots \qquad \vdots$$
$$F_{16} = 15_{10}$$

where the subscript indicates the value of the radix.

2.2.4   Conversion from Decimal to Hexadecimal

The process is similar to that of converting decimal to binary:-

$$123_{10}/16_{10} = 7_{10} \quad \text{remainder } 11_{10} = B_{16}$$
$$7_{10}/16_{10} = 0_{10} \quad \text{remainder } 7_{10} = 7_{16}$$

Reading the remainder column from the bottom gives the converted value.

$$123_{10} = 7B_{16}$$

2.2.5   Binary-Hex conversion

Each hexadecimal digit may be represented by 4 binary digits, since $2^4 = 16$. A table is given below:-

Hexadecimal	Binary
0	0000
1	0001
2	0010
3	0011
4	0100
5	0101
6	0110
7	0111
8	1000
9	1001
A	1010
B	1011
C	1100
D	1101
E	1110
F	1111

The convenience of hexadecimal numbers is that we can translate easily between "hex" and binary, simply by replacing each hex digit by its corresponding (4 digit) binary equivalent. Thus if we want to examine a number held within a computer we can easily expand the 'hex' to give the individual binary digits, yet we can print the values in hex to save space.

It is also important to note that the basic unit in terms of electronic data transfer is the 'byte' which consists of 8 binary bits or 2 hex characters. A byte can therefore be expressed as a two digit hex number.

Exercises

1. Represent the following numbers in scientific notation with a mantissa in the range 1 to 9.$\dot{9}$

   (a) 12.63

   (b) 8000000

   (c) $10\frac{1}{2}$

   (d) $3\frac{1}{3}$

   (e) 1/100

2. Complete the following table of binary, decimal and hexadecimal numbers.

Binary	Decimal	Hex
1011		
	13	
		$42A$

3. What is the result of dividing the hexadecimal number $17DDE$ by 2?

Answers

1. (a) 1.263×10^1 or $1.263E1$ or $1.263_{10}1$

   (b) 8×10^6

   (c) 1.05×10^1

   (d) 3.333333×10^0

   (e) 1×10^{-2}

2. The completed table is:-

Binary	Decimal	Hex
1011	11	B
1101	13	D
010000101010	1066	$42A$

3. The result of dividing the hexadecimal number $17DDE$ by 2 is BEEF.

Chapter 3

Powers and Logarithms

The main reason for this chapter is to show you that difficult formulæ (which you have been taught to remember) are not thought up in isolation. They are usually derived from a few simple assumptions, from which they follow logically. Once you have followed the logic, life becomes easier because you understand why the formulae work. Not only that, if pushed you can derive them again yourself without having to fill your head with unnecessary details. Memory is no substitute for understanding, it helps - but you can always look things up if you know where to search.

Your parents/grandparents hated logarithms ("logs") - because performing calculations using log tables was difficult and tedious. Now for the good news; no-one of sound mind does this any more! However, the manipulation of expressions involving powers is an essential skill if you are to deal with scientific calculations. Having gained that skill, understanding logarithms follows naturally.

Now read on: understanding powers and logarithms is good, doing arithmetic with logs is a waste of time, calculators do it better!

3.1 Powers and Indices

We define a^2 as the number resulting when 2 copies of a are multiplied together.

$$a^2 = a \times a$$

Similarly

$$a^3 = a \times a \times a$$

Now for the difficult bit. We could go on defining $a^4, a^5 \cdots$ but this is a pain, so we think about the general case - $a^{anything}$. We define a^m, where m is any number, as the number resulting when m copies of a are all multiplied together.

$$a^m = a \times a \times a \times \cdots \times a$$

m is called the *power* or *index* and a is referred to as the *base*. In an equation like the one above where several things are to be multiplied together the individual terms are referred to as *factors*. We refer to the expression a^m as a to the power m, or "a to the m" for short. Expressions like a^m, 3^2 and 10^{-13} are called exponential expressions.

3.1.1   Some general rules of powers and indices

We shall restrict ourselves to the case where a is any real number other than zero and m and n are integers, though the following results can be shown to apply more generally.

$a^m \times a^n$

$$a^m = \underbrace{a \times a \times \cdots \times a}_{m \text{ factors}}$$

$$a^n = \underbrace{a \times a \times \cdots \times a}_{n \text{ factors}}$$

$$\therefore a^m \times a^n = \underbrace{a \times a \times \cdots \times a}_{m+n \text{ factors}}$$

$$\therefore a^m \times a^n = a^{m+n}$$

e.g. $\qquad a^2 \times a^3 = a^{2+3}$
$$= a^5$$

e.g. $\qquad 10^5 \times 10^2 = 10^7$

a^0

$$a^m \times a^0 = a^{m+0} = a^m$$

$$\therefore a^0 = 1$$

because multiplying by a^0 leaves any value unchanged.

a^{-m}

$$a^m \times a^{-m} = a^{m-m} = a^0 = 1$$

$$\therefore a^{-m} = \frac{1}{a^m}$$

a^m/a^n

$$\frac{a^m}{a^n} = \frac{a \times a \times \cdots a}{a \times a \times \cdots a} \qquad \begin{matrix} m \text{ factors} \\ n \text{ factors} \end{matrix}$$

$$= a \times a \times \cdots \times a \qquad (m-n) \text{ factors}$$

$$\therefore \frac{a^m}{a^n} = a^{m-n}$$

$(a^m)^n$

$$
\begin{aligned}
(a^m)^n &= a^m \times a^m \times \cdots \times a^m \quad (n \text{ factors}) \\
&= a^{m+m+\cdots+m} \\
\therefore (a^m)^n &= a^{m\times n}
\end{aligned}
$$

e.g. $\qquad (10^2)^3 \;=\; 10^6$

$a^{1/m}$

$$\text{Consider } a^{1/m} \times a^{1/m} \times \cdots \times a^{1/m} \quad (m \text{ factors})$$

$$= a^{(1/m+1/m+\cdots+1/m)} = a^1 = a$$

$$(a^{1/m})^m = a$$

$$\therefore a^{1/m} = \sqrt[m]{a}$$

3.1.2 Rules of Powers and Indices - Summary

$$a^m = a \times a \times a \times \cdots \times a$$

$$a^m \times a^n = a^{m+n}$$

$$a^0 = 1$$

$$a^{-m} = \frac{1}{a^m}$$

$$\frac{a^m}{a^n} = a^{m-n}$$

$$(a^m)^n = a^{m\times n}$$

$$a^{1/m} = \sqrt[m]{a}$$

3.2   Logarithms

In the "olden days" all arithmetic was done using pencil, paper and the stuff which keeps your ears apart. This was hard, and therefore any tool or magic spell which could facilitate arithmetic was perceived to be a very good thing!

Logarithms were invented by a Scotsman named John Napier - who probably had nothing better to do during the long winter nights, other than chuck the odd log under his still. He realised that by representing numbers as powers of 10, and then using the rules derived in the last section, he could avoid a lot of tedious arithmetic (idle mathematicians again!)

His idea was to produce a look-up table relating x and p where $x = 10^p$ (log tables) and a reverse or antilog table which would allow the user to find x given p. Then, in order to calculate $x \times y$:

1. look up the logarithm of x (p) and the logarithm of y (q).

2. calculate $r = p + q$

3. look up the antilog of r to give z, where $z = x \times y$.

e.g. to calculate 45.67×23.45

1. look up $\log 45.67 = 1.6596$ and $\log 23.45 = 1.3701$

2. add them together to give 3.0297

3. look up the antilog of 3.0297 which gives the answer 1071 to 4 significant figures.

Thus, long multiplication was reduced to looking up three values in tables, together with one addition. This saved a <u>lot</u> of work and reduced the chances of error. Further exploitation of the properties of exponentials allow the simplification of other arithmetic calculations.

This was important because rapid calculations, together with accurate timing, were essential to navigation. Good navigation, in turn was the key to global power. (Actually it meant that entrepreneurs could pilfer all manner of treasure throughout the world and then find their way back!)

Until the 1970's every science student had a book of log-tables, though some people, usually engineers, had fancy gadgets called slide rules which were a sort of automated set of log tables. Mathematicians and *"real scientists"* referred to them as "guessing sticks". The equivalent of *log* and *antilog* tables are the "log" and "10^x" buttons on your calculator. Nowadays log tables are confined to antiquarian bookshops (or the fire-back!) as a result of the calculator revolution. In fact the electronic pocket calculator still uses logarithms, but does so largely in secret. Sometimes you may notice strange numbers appear whilst you are performing calculations - these are probably logarithms used by your calculator during the intermediate steps of a calculation.

To a large extent the overt use of logarithms can now be avoided because of the power of the computer. However there are still many applications and protocols which make use of the old fashioned ways and so it is both useful and informative to have a basic understanding of how logarithms work.

3.2.1 What are logarithms?

We shall start by discussing logarithms to base 10 because they are widely used. However, it should be born in mind that we could use logs to any base, though the only other base of any consequence is that of e, the exponential constant $2.718282\ldots$

Basically $\log(x)$ is the number of 10's which you have to multiply together to generate the value x. Thus

$$\log(100) = 2 \quad \text{because} \quad 100 = 10 \times 10$$

In the same way

$$\log(1000000) = 6 \quad \text{because} \quad 1000000 = 10 \times 10 \times 10 \times 10 \times 10 \times 10$$

So far this is easy! but you (and I) have difficulty in understanding $\log(50)$. Suffice it to say that it is bigger than 1 because $10^1 = 10$ and less than 2 because $10^2 = 100$ (actually it is 1.6990 to 4 decimal places). Incidentally if you calculate $\log(500)$ you will notice that it is exactly 1 bigger than $\log(50)$ *ie* 2.6990. This is because you need to multiply by one more ten to get 500 than you do to get 50! ($500 = 50 \times 10$) Likewise $\log(5000)$ is one more than $\log(500)$. Notice that numbers formed by repeatedly multiplying by 10 have logarithms which increase by 1 at each multiplication.

Repeated multiplication by any number produces a sequence of values whose logarithms behave in the same way, *ie* their log values increase by a constant amount at each multiplication. Consider the sequence 5, 25, 125... and the corresponding log values:

x	5	25	125	625
$\log(x)$	0.6990	1.3979	2.0969	2.7959

Notice that as the number is multiplied at each step by 5, the corresponding logarithm is increased by 0.6990. (Because $\log(5) = 0.6990$)

Many physical relationships involve repeated multiplication: e.g. growth of a population over successive generations, compound interest, radioactive decay, repeated dilutions, and the frequency of successive notes in the music scale. Plotting a simple graph of the logarithm of these values against time should reveal a straight line relationship because the logarithm will increase (or decrease) at a constant rate.

3.2.2 Definition

The value of a logarithm depends upon two things; the number itself, and the *base* of the logarithms.

The logarithm of N to the base a is usually written $\log_a N$, and is defined as follows:

$$\text{if}\quad a^x = N \quad\text{then}\quad \log_a N = x \tag{3.1}$$

It is "the power to which the base has to be raised in order to generate the number". (i.e. how many a's do we need to multiply together to produce the value N) So that

$$N = a^{\log_a N} \quad \text{(obviously!?)}$$

For example, the logarithm to base 3 of 9 ($\log_3 9$) is 2 because $9 = 3^2$.

$$\text{Examples:-}\qquad
\begin{aligned}
2^4 &= 16 & \therefore\ \log_2 16 &= 4\\
12^2 &= 144 & \therefore\ \log_{12} 144 &= 2
\end{aligned}$$

Common Logarithms

Traditionally, tables of logarithms to the base 10 were used, and these were known as *common logarithms.*

Natural Logarithms

Mathematicians and physicists tend to use logarithms to the base e; e is the number 2.718282... which is sometimes referred to as Euler's number or the "exponential constant". This may seem strange but the number e occurs naturally in many areas of science. Logarithms to base e are referred to as *natural logarithms.* The modern equivalent of natural log and antilog tables are the ln and e^x buttons on your calculator.

The exponential function $exp(x)$ returns the value of e^x so that the two expressions are equivalent.

Logarithms to base 10 and e occur so often that they are abbreviated as follows:

$$\begin{aligned}
\log(x) &\equiv \log_{10}(x)\\
\ln(x) &\equiv \log_e(x)
\end{aligned}$$

Generally you are advised to work with natural logarithms because the algebra is easier, though sometimes (eg when calculating pH) you will find it more convenient to work to base 10.

e - an Interesting Number

Suppose that your bank suggested an investment for a specified period, at the end of which your investment would be returned together with 100% interest. Thus if you invested £1, at the end of the period you would collect £2.

If you were able to persuade the bank to pay you at half the rate, but compounded over two half periods, your final return would be:

$$(1+1/2) * (1+1/2) = (1+1/2)^2 = £2.25$$

since you would receive 50p interest half way through the period, and then you would have £1.50 invested over the remaining half period.

You may even persuade the bank to give you interest every quarter, in which case you would receive:

$$(1+1/4) * (1+1/4) * (1+1/4) * (1+1/4) = (1+1/4)^4 = £2.44$$

Being a thinking person you would obviously see advantage in taking this further. A general formula when the period is split into n equal intervals is:

$$(1+1/n)^n$$

from which you could calculate the final values.

n	final value
1	2.00
2	2.25
4	2.44
10	2.59
100	2.70

It is clear that as the value of n increases, the result of evaluating $(1+1/n)^n$ approaches some limiting value. As mathematicians we write this as follows:

$$Lt_{n\to\infty}(1+1/n)^n = 2.718281828\ldots$$

by which we mean - the limiting value of this expression, if we were able to calculate it with n taking an infinite value (which we can't!), would be $2.718281828\ldots$ We represent this limiting value as e and refer to it as "Euler's number" or simply "e".

3.2.3   Mathematical Derivation of the Rules of Logarithms

In order to demonstrate that the following rules are independent of the base chosen we shall work with logarithms to the base a, where a is any positive number. These rules are closely related to those for *powers*.

$\log_a 1$

$$\log_a 1 = 0 \quad \text{because} \quad a^0 = 1$$
$$ln(1) = 0 \quad \text{because} \quad e^0 = 1$$
$$log(1) = 0 \quad \text{because} \quad 10^0 = 1$$

$\log_a a$

$$\log_a a = 1 \quad \text{since} \quad a^1 = a$$

$\log_a(M \times N)$

$$\text{Let} \quad x = \log_a M \quad \therefore a^x = M$$
$$\text{and} \quad y = \log_a N \quad \therefore a^y = N$$

$$M \times N = a^x \times a^y = a^{x+y}$$
$$\log_a(M \times N) = x + y$$
$$\log_a(M \times N) = \log_a M + \log_a N$$

$\log(M/N)$

$$\text{Let} \quad x = \log_a M \quad \therefore a^x = M$$
$$\text{and} \quad y = \log_a N \quad \therefore a^y = N$$

$$\frac{M}{N} = \frac{a^x}{a^y} = a^{x-y}$$
$$\log_a(M/N) = x - y$$
$$\log_a(M/N) = \log_a M - \log_a N$$

$\log_a(M^p)$

$$\text{let} \quad x = \log_a M$$
$$\therefore M = a^x$$
$$\therefore M^p = a^{px}$$
$$\therefore \log_a(M^p) = px$$
$$\log_a(M^p) = p \log_a M$$

Example - Calculate $\log_{10}(10 \times 100)$

$$\log_{10}(10 \times 100) = \log_{10} 10 + \log_{10} 100 = 1 + 2 = 3$$

Example - Calculate $\log_2(2 \times 16)$

$$\log_2(2 \times 16) = \log_2 2 + \log_2 16$$
$$= 1.0 + 4.0$$
$$= 5 \qquad (= \log_2(32))$$

Example - Calculate $\log_4(16^3)$

$$
\begin{aligned}
\log_4(16^3) &= 3 \times \log_4 16 \\
&= 6 \qquad (= \log_4 4096)
\end{aligned}
$$

Example - Find the logarithm of $32\sqrt[5]{4}$ **to base** $2\sqrt{2}$

Let x be the required logarithm

$$
\begin{aligned}
\text{then } (2\sqrt{2})^x &= 32\sqrt[5]{4} \\
(2.2^{\frac{1}{2}})^x &= 2^5.2^{\frac{2}{5}} \\
2^{\frac{3}{2}x} &= 2^{5+\frac{2}{5}} \\
\frac{3}{2}x &= \frac{27}{5} \\
x &= \frac{18}{5} = 3.6 \\
\log_{2\sqrt{2}}(32\sqrt[5]{4}) &= 3.6
\end{aligned}
$$

3.2.4   Calculating logarithms to a different base

Suppose we require $\log_b N$ having been given $\log_a N$.

$$
\text{let } \log_b N = y \text{ so that } b^y = N
$$

$$
\therefore \ \log_a N = \log_a(b^y)
$$

$$
\therefore \ \log_a N = y \log_a b
$$

$$
\therefore \ y = \frac{\log_a N}{\log_a b}
$$

$$
\log_b N = \frac{\log_a N}{\log_a b} \tag{3.2}
$$

Example - Suppose we know $\ln(2)$ **but need** $\log(2)$.

$$
\log_{10} 2 = \frac{\log_e 2}{\log_e 10} = \frac{0.6931}{2.3026} = 0.3010
$$

3.2.5   Rules of Logarithms - Summary

$$\log_a 1 = 0$$

$$\log_a a = 1$$

$$\log_b X = \frac{\log_a X}{\log_a b}$$

$$\log_a XY = \log_a X + \log_a Y$$

$$\log_a \frac{X}{Y} = \log_a X - \log_a Y$$

$$\log_a X^n = n \log_a X$$

The above relationships apply whatever the base (a) of the logarithms.

3.3   Exponential Functions

Population Dynamics

The usual basis for the description of population growth is the exponential equation

$$N_t = N_0 \, e^{rt}$$

in which N_0 is the number of individuals in the initial population and r is known as the "intrinsic rate of natural increase of the population". This form of the equation does not lend itself to understanding the basics of population growth. The difficulty is mainly due to the inclusion of the exponential constant e, which is a strange number to those other than mathematicians.

However, if we use the relationship

$$x^{pq} = (x^p)^q$$

we can see that the above expression may be represented as follows:

$$N_t = N_0 R^t$$

where

$$R = e^r$$

In particular, if we now refer to R as the intrinsic rate of increase, R has a much more intuitive meaning; since when R is greater than 1, the population is increasing - while when R is less than 1 the population will decrease. Furthermore, when $R=1$ the population is stable.

Also this expression allows us to see that as t increases by a unit value, the population number will increase by the factor R:

$$N_{t+1} = N_t R$$

since

$$N_0 R^{t+1} = N_0 R^t R$$

Exercises

1. Evaluate the following expressions using powers of thought and pen or pencil only.

   (a) $16^{\frac{1}{2}}$

   (b) $27^{\frac{2}{3}}$

   (c) $\log 1$

   (d) $\ln e^7$

   (e) $\log_9 3$

   (f) $\log \frac{1}{10000}$

   (g) $\log_2(\frac{1}{8})$

   (h) $\log_q q^5$

   (i) $\log_2 16$

2. Prove:- $\log_a b \times \log_b c \times \log_c a = 1$

3. If $6^{3x} = 14.7$ calculate x.

4. The pH of a chemical solution is defined as follows:-

$$pH = -\log_{10}(\text{hydrogen ion concentration})$$

   What is the hydrogen ion concentration of a solution with $pH = 4.2$?

5. Measurements of the concentration C of a substance are to be taken at several values of time t after the start of an experiment. It is expected that the values of C can be predicted by the equation

$$C = C_0 e^{-kt}$$

   where C_0 is the initial concentration, t is the time and k is a constant who's value is sought. Estimate the value of k from the following data by plotting $\ln C$ against t:

t	0	5	10	15	20
C	20	12	7.3	4.5	2.7

Answers

1. (a) 4

   (b) 9

   (c) 0

   (d) 7

   (e) $\frac{1}{2}$

   (f) -4

   (g) -3

   (h) 5

   (i) 4

2.

$$\log_a b \times \log_b c \times \log_c a$$
$$= \log_a b \times \frac{\log_a c}{\log_a b} \times \frac{\log_a a}{\log_a c}$$
$$= \log_a a$$
$$= 1$$

3.

$$6^{3x} = 14.7$$
$$\therefore \ 3x \log 6 = \log 14.7$$
$$x = \frac{\log 14.7}{\log 6} \times \frac{1}{3}$$

4. Let H^+ be the hydrogen ion concentration.

$$4.2 = -\log H^+$$
$$-4.2 = \log H^+$$
$$H^+ = 10^{-4.2}$$
$$H^+ = 6.310 \times 10^{-5}$$

5. Taking logarithms of both sides of the equation we have:

$$\ln C = \ln C_0 - kt$$

Plotting $\ln C$ against t we obtain a straight line whose slope is -0.1. Hence $k = 0.1$

Chapter 4

Calculations and Applications

4.1 Convert miles/hour (mph or miles hour^{-1}) to m s^{-1}

The way to tackle unit conversion problems is to work through, changing one unit at a time, in successive stages as follows:

$$\frac{miles}{hour} \times 1.609 \Rightarrow \frac{km}{hour} \times 1000 \Rightarrow \frac{m}{hour} \times \frac{1}{60 \times 60} \Rightarrow \frac{m}{s}$$

and combining all these factors produces the formula

$$\frac{miles}{hour} \times \frac{1.609 \times 1000}{60 \times 60} \Rightarrow \frac{m}{s}$$

A "back of the envelope calculation" gives:

$$\frac{1.609 \times 1000}{60 \times 60} = \frac{1.609 \times 10}{6 \times 6} = \frac{16.09}{36} \approx 0.5$$

which the calculator confirms to give:

$$\frac{miles}{hour} \times 0.447 \Rightarrow \frac{m}{s}$$

and hence to convert a speed in miles/hour to m s^{-1} we multiply by 0.447 (to 3 significant figures).

4.2 The pH of a solution

Definition - the pH of a solution is the negative of the logarithm to base 10 of the hydrogen ion activity.

If a biological fluid has a pH of 7.5, what is the hydrogen ion activity?

$$7.5 = -\log_{10}(H^+)$$
$$\therefore H^+ = 10^{-7.5} \quad \text{from the definition of logarithm (\textit{equation} (3.1))}$$
$$= 3.162 \times 10^{-8}$$

4.3   How many microbes? The Viable Count Method

In this method different concentrations of microbes are made up by successively diluting by a constant factor. A small sample from each dilution is plated and counts are made of the cfu's (colony forming units) present in each of the samples. The plate containing between 30 and 300 cfu's is normally regarded as the sample to be used in calculating the original concentration. If, for this sample:

n is the number of cfu's counted

d is the number of dilutions for this plate

v is the volume of sample plated (ml)

$f > 1$ is the dilution factor, the concentration of each solution is $1/f$ times the previous concentration. (f is commonly 10).

The concentration c of cells in the original sample can be calculated as follows:

$$c = \left(\frac{n}{v}\right) \times f^d \text{ ml}^{-1}$$

The term n/v gives the concentration (ml^{-1}) of cfu's in the plated sample, while the term f^d scales the count according to the number of dilutions.

Thus if we counted 40 cfu's on the plate which had 5 dilutions of factor 10, and the plate sample volume was .05 ml the calculation would be

$$c = (40/.05) \times 10^5 = 8 \times 10^7 \text{ ml}^{-1}$$

Note 50 $\mu l = .05$ ml

4.4   Surface Area of Humans

Dubois and Dubois [1] showed that the surface area of the human body may be approximated by the formula:

$$S = 0.007184 W^{0.425} H^{0.725} \tag{4.1}$$

---

[1]D. Dubois & E. F. Dubois: *A formula to estimate the approximate surface area if height and weight be known.* Archives of Internal Medicine, Chicago, 1916, 17: 863-871.

where: S = surface area (m^2), W = weight (kg), and H = height (cm). Thus the surface area of a person of height 170cm and weight 70kg is predicted to be 1.810m^2.

Many older people (e.g. professors) and Americans would find this formula extremely difficult. They would prefer to work in terms of feet, pounds and inches so we may be asked (told!) to provide a simpler version.

If we are to use *equation* (4.1) with inputs in pounds and inches we must first of all convert these values into the units expected by the formula, so that if w is the weight in pounds and h the height in inches we have:

$$
\begin{aligned}
W &= \frac{w}{2.2046} \\
H &= 2.540 \times h
\end{aligned}
$$

since 1 kilogram $\approx$ 2.2046 pounds and 1 inch $\approx$ 2.540 cm. Substitution in *equation* (4.1) gives:

$$S = 0.007184 \left(\frac{w}{2.2046}\right)^{0.425} (2.540 \times h)^{0.725} \qquad (4.2)$$

where the input values w and h are now in the required units, but the output is still calculated in terms of square metres. In order to calculate s, the area in square feet we must introduce the additional conversion:

$$S = \frac{s}{3.2808^2}$$

since 1 metre $\approx$ 3.2808 feet and hence 1 square metre is 3.2808 $\times$ 3.2808 square feet.

Substituting this in *equation* (4.2) gives:

$$\frac{s}{3.2808^2} = 0.007184 \left(\frac{w}{2.2046}\right)^{0.425} (2.540 \times h)^{0.725} \qquad (4.3)$$

where all the values are now in the required units. This is extremely messy so here is the tidying up in stages:

$$
\begin{aligned}
s &= 3.2808^2 \times 0.007184 \times \left(\frac{w}{2.2046}\right)^{0.425} \times (2.540h)^{0.725} \\
&= 3.2808^2 \times 0.007184 \times \left(\frac{1}{2.2046^{0.425}} \times w^{0.425}\right) \times (2.540^{0.725} \times h^{0.725}) \\
&= \left[\frac{3.2808^2 \times 0.007184 \times 2.540^{0.725}}{2.2046^{0.425}}\right] \times w^{0.425} \times h^{0.725}
\end{aligned}
$$

Evaluation gives the required working version of *equation* (4.1):

$$s = 0.1086 w^{0.425} h^{0.725}$$

Notice that the powers of the weight and height terms remain unaltered - only the initial constant has changed.

4.5   Fluid Flow - Poiseuille's Formula

The velocity of flow, or volume flux, of a fluid along a cylindrical vessel of length l and radius a is given by

$$V = \frac{\pi(\Delta_p)a^4}{8\eta l} \quad \text{Poiseuille's formula}$$

where η is a constant characteristic of the fluid (viscosity) and Δ_p is the pressure difference between the ends of the pipe.

This equation has applications in the study of blood flow, food processing, coolant, lubricants etc.

Suppose we reduce the radius of the tube to half its original value, what effect does this have on V?

$$
\begin{aligned}
V^* &= \frac{\pi(\Delta_p)(a/2)^4}{8\eta l} \\
&= \frac{\pi(\Delta_p)a^4}{8\eta l.2^4} \\
&= \frac{1}{16}V
\end{aligned}
$$

The reduction of the radius to one half reduces the volume flux by a factor of 16. One consequence of this is that your heart must work 16 times harder in order to deliver the same amount of blood if your arteries are reduced in diameter by one half!

4.6   The Growth of a Bacterial Population

If we assume that the population is not restricted by space, lack of nutrients etc., the population will double at each generation. Thus if we have N_0 bacteria at time 0, there will be $N_0 \times 2$ bacteria after 1 generation period.

$$N_1 = N_0 \times 2$$

Similarly during the next generation period,

$$
\begin{aligned}
N_2 &= N_1 \times 2 \\
&= N_0 \times 2 \times 2 \\
&= N_0 \times 2^2
\end{aligned}
$$

After t such time intervals the number of bacteria in the population is given by

$$N_t = N_0 \times 2^t$$

For *E. coli* the generation time is 20 mins so that after 20 generations (6 hours 40 minutes) we can predict the population as follows:

$$
\begin{aligned}
N_{20} &= N_0 \times 2^{20} \\
&= N_0 \times 2^{10} \times 2^{10} \\
&\qquad (2^{10} = 1024) \\
&\simeq N_0 \times 1 \; million
\end{aligned}
$$

4.7 The Beer-Lambert Law

Beers Law states that when light passes through a solution the intensity of the emergent light (I) is less than that of the incident light (I_0). The relationship is given by:

$$I = I_0 10^{-\varepsilon cd}$$

where

c is the concentration of the solution (moles/litre)

d is the length of the light path through the liquid

ε is the extinction coefficient

The above equation allows us to measure concentrations using colorimetry as follows:

1. Measure the intensity I_s for a standard solution of concentration c_s to give:

$$I_s = I_0 \; 10^{-\varepsilon c_s d} \tag{4.4}$$

2. Measure the intensity I_t of the test solution whose concentration c_t is unknown to give

$$I_t = I_0 10^{-\varepsilon c_t d} \tag{4.5}$$

Take logs (to base 10) of equations 4.4 and 4.5 and rearrange each equation to give

$$
\begin{aligned}
c_s &= \frac{\log I_0 - \log I_s}{\varepsilon d} \\
c_t &= \frac{\log I_0 - \log I_t}{\varepsilon d} \\
\therefore \frac{c_t}{c_s} &= \frac{\log I_0 - \log I_t}{\log I_0 - \log I_s} \\
\therefore c_t &= c_s \frac{\log(I_0/I_t)}{\log(I_0/I_s)} \tag{4.6}
\end{aligned}
$$

4.8   A River Pollution Incident

An organisation has been identified to have polluted a lake by dumping a tank full of "nasty stuff" into it!

The river authority know exactly when the stuff was dumped and have measured the concentration as 10ppm 10 hours after the incident and 7ppm after 13 hours.

The organisation claim that the initial concentration was only 20ppm which is below the EEC limit of 25. Do we believe them?

We assume that the concentration can be described by an exponential decay relationship as follows:

$$C = C_0 e^{-kt}$$

where C is the concentration (ppm) at time t, C_0 is the original concentration, e is the exponential constant and k is a rate constant with units of $time^{-1}$ which depends upon the mixing and flow rate through the lake.

We need to find the two constants C_0 and k. In order to do this we use the two measurements that have been taken to generate two equations in the two unknowns as follows:

$$10 \ = \ C_0 e^{-10k} \tag{4.7}$$
$$7 \ = \ C_0 e^{-13k} \tag{4.8}$$

To simplify these equations we take logs to base e.

$$\ln 10 \ = \ \ln C_0 - 10k \tag{4.9}$$
$$\ln 7 \ = \ \ln C_0 - 13k \tag{4.10}$$

Note that $\ln(e^z) = z$ from the definition of logarithm (*equation* (3.1)) - think about it!

Subtracting *equation* (4.10) from *equation* (4.9) we have

$$\ln 10 - \ln 7 \ = \ -10k - (-13k)$$

and hence

$$k \ = \ \frac{\ln 10 - \ln 7}{3} = \frac{\ln(10/7)}{3} \tag{4.11}$$

Now we return to *equation* (4.7) in order to find the value of C_0 as follows:

$$10 = C_0 e^{-10k}$$

Multiply both sides by e^{10k} to give

$$10 e^{10k} = C_0 e^{-10k} e^{10k} = C_0 e^0 = C_0$$

Reversing the order and substituting k from *equation* (4.11) we have

$$
\begin{aligned}
C_0 &= 10e^{10(\frac{\ln(10/7)}{3})} \\
&= 32.8 \text{ ppm}
\end{aligned}
$$

At which point we take the offender to the EEC cleaners! *ie* the original input was above the regulation threshold.

4.9 Linear Regression

4.9.1 Notation for sums of sequences

Mathematicians often work with sums of values. As you would expect they have developed a notation to avoid writing down all the individual values. For example, if it was necessary to refer to the sum of the first n integer numbers, this would be expressed as

$$
\sum_{i=1}^{n} i
$$

which is mathematical shorthand for

$$
1 + 2 + 3 + \ldots + n
$$

Notice that we define the starting and finishing values below and above the Σ character and define the general term to be summed in an expression following it. This expression can be as complicated as necessary - the following defines the sum of the squares of all the even numbers from 2 to 100:

$$
\sum_{i=1}^{50} (2i)^2
$$

If we have a set of data - $(x_1, y_1), (x_2, y_2), \ldots (x_n, y_n)$ we could expres the sum of the individual x values and the sum of the xy products as

$$
\sum_{1}^{n} x_i \text{ and } \sum_{1}^{n} x_i y_i
$$

respectively. However if we intend the sum to include all the values, as in the case above, it is not necessary to specify the starting and finishing values. The following equation defines the mean of the x values:

$$
mean = \frac{\sum x}{n}
$$

4.9.2   Fitting the best Straight Line

The values of x and y in the table below were measured during an experiment and are expected to be linearly related by the formula $(y = mx + c)$ where the parameters m and c are to be determined. The best fit is calculated by minimising the following sum of squares

$$SS = \sum_{i=1}^{n} [y_i - mx_i + c]^2$$

where y_i are the n observed values, $n = 6$ in this example, and $mx_i + c$ are the values predicted from the parameters m amd c. It can be seen that when the fit is good SS will be small and a large value of SS indicates a poor fit. In the case of a straight line relationship it is possible, using methods of calculus, to generate formulae for the best values of the two parameters (i.e. that minimise SS)as follows:

$$m = \frac{\Sigma xy - \Sigma x \Sigma y/n}{\Sigma x^2 - \Sigma x \Sigma x/n}$$

$$c = \Sigma y - m \Sigma x/n$$

where n is the number of data points. This process is known as linear regression using the method of least squares.

x	0	1	2	3	4	5
y	0.9	3.2	4.8	7.0	8.7	11.1

We can perform the calculation by setting up a table as follows: (on many pocket calculators/computers there will be no need to do this since the calculation can be done automatically)

	x	y	x^2	xy
	0	0.9	0	0
	1	3.2	1	3.2
	2	4.8	4	9.6
	3	7.0	9	21.0
	4	8.7	16	34.8
	5	11.1	25	55.5
Totals	15	35.7	55	124.1

slope

$$m = \frac{124.1 - (15)(35.7)/6}{55 - (15)(15)/6} = 1.986$$

intercept

$$c = 35.7/6 - 1.986(15)/6 = 0.995$$

The equation relating x and y is therefore

$$y = 1.986x + 0.995$$

4.10 The Michaelis-Menton equation

The Michaelis-Menton equation predicts the reaction velocity V in terms of the substrate concentration $[S]$ as follows:

$$V = \frac{V_{max}[S]}{[S] + K_m} \qquad (4.12)$$

where V_{max} and K_m are parameters specific to the reaction, and which usually are determined from experiment and subsequent function fitting.

Given a table of data values $([S]_i, V_i)$ the problem is to define the values of the parameters V_{max} and K_m.

In the "olden days" this was difficult, because there is no analytical solution which will generate the best values. Fitting curves was either achieved "by eye" or by transforming the data in some way so that the transformed data could be fitted. This usually implied that the transformation should result in a linear relationship so that the least squares fit (See section 4.9) could be used.

Scientists went to great lengths in order to linearise their data. In the case of fitting the Michaelis-Menton equation two alternative transformations have been used.

4.10.1 The Lineweaver-Burke transformation

We begin by multiplying the Michaelis - Menton equation (*equation* (4.12)) by $([S] + K_m)$ to give

$$([S] + K_m)V = V_{max}[S] \qquad (4.13)$$

divide by $[S]$

$$\left(1 + \frac{K_m}{[S]}\right)V = V_{max}$$

divide by V and invert the equation

$$\frac{V_{max}}{V} = 1 + \frac{K_m}{[S]}$$

and finally, dividing by V_{max} gives the Lineweaver-Burke equation:

$$\frac{1}{V} = \frac{1}{V_{max}} + \frac{K_m}{V_{max}} \times \frac{1}{[S]} \qquad (4.14)$$

If we let $y = 1/V$ and $x = 1/[S]$ this is the equation of a straight line whose intercept is $1/V_{max}$ and slope is K_m/V_{max}. Thus we can find the values of K_m and V_{max} by fitting the best (least squares) line.

4.10.2   The Eadie-Hofsee transformation

Here we multiply out the RHS of *equation* (4.13)

$$V[S] + VK_m = V_{max}[S]$$

divide through by $[S]$

$$V + \frac{VK_m}{[S]} = V_{max}$$

and re-arrange to give the Eadie-Hofsee equation

$$V = V_{max} - K_m \left(\frac{V}{[S]} \right) \qquad (4.15)$$

Plotting V against $V/[S]$ should give a straight line of slope $-K_m$ and intercept V_{max}.

4.10.3   Fitting the parameters the Modern Way

There is no analytical formula for the best (least squares) parameter values, but we can calculate the sum of squares of the differences between the observed data and the values predicted by the equation using estimates for the values of V_{max} and K_m.

$$SS = \Sigma \left(V_i - \frac{V_{max}[S]_i}{[S]_i + K_m} \right)^2$$

It is then possible to optimise the fit by changing the values of the parameters so that the value of SS is minimised.

Initial estimates of the parameters V_{max} and K_m can be found as follows:

1. An estimate of V_{max} is easily gained from the data set by taking the maximum observed reaction velocity.

2. Estimating K_m is a little more difficult, but consideration of the equation provides a useful pointer as follows:

$$V = \frac{V_{max}[S]}{[S] + K_m}$$

$$\therefore K_m = \frac{V_{max}}{V}[S] - [S]$$

$$= [S] \left(\frac{V_{max}}{V} - 1 \right)$$

from this equation we can see that when $V = V_{max}/2$, $K_m = [S]$, so that we can estimate K_m from the table by finding the value of $[S]$ when $V \approx V_{max}/2$.

The calculation is best done using a computer spreadsheet. An initial plot (X-Y) of the experimental data will show up outlier points which should be checked. It will also be beneficial to plot the observed and predicted values and to try different parameter values in order to see how the fitted equation reacts. Progress can be made by manually changing the parameters to reduce the sum of squares value. This will be a time consuming process, but it is possible on most computer systems to optimise the fit automatically.

There are several advantages of this modern calculation:

- no transformation is necessary.

- plotting the curve gives added insight, especially when the parameter values are changed manually. The shape of the curve and its response to changes in parameter values can be seen.

- the calculation is easier.

- the calculations are carried out on the raw (untransformed) data so that the form of the errors is known.

- the errors on individual points can easily be weighted if necessary.

4.11 Graphs and Functions

It is almost always useful to picture data and functions in graphical form. Personal computers make life easy in this respect because data can be collected and displayed in spreadsheet form. You should **always** do this: it is much easier to detect anomalies from a graph than it is from a table, and it is always helpful to see the shape of your data.

4.11.1 Plotting Graphs

It is rarely necessary for a scientist to need anything other than an $x-y$ plot. Moreover, the initial look should be at an $x-y$ plot with no transformations and no lines joining the points. The reason for this is that transformations destroy the "shape" of the data, and drawing smooth lines joining experimental points can be misleading - there is no reason to believe that the fitted lines are an accurate representation of data that was not collected. Sometimes, if there are different data sets on the same graph, it is helpful to join the points within each set with straight lines so that the sets can be distinguished.

Spreadsheet graphics is a subject in itself and one which occupies more time and effort than it should! Remember that the facilities provided in the popular packages were designed for salespersons to impress their customers/supervisors, they were not designed particularly to help scientists. Most of the facilities provided are of little relevance to us.

After the data has been plotted and checked (*and a backup copy saved*), then is the time to analyse the underlying trends and relationships. It may be that the relationships are known, in which case you should calculate the best fit to the data as in section (4.10.3). If the relationship is not known, consideration of the initial plot may reveal a shape that is familiar. Try fitting it, you may make discoveries, and even if you don't you will learn a lot about the data. And it can be both rewarding and fun. If the data-set is very noisy, even this may not be possible and a statistical investigation may prove to be the best way forward. Whatever the approach and subsequent analysis, it is our job as scientists to provide justification for any conclusions that we make. This will normally take the form of a statistical analysis, unless the relationship is clearly obvious.

4.11.2   Shapes of some useful functions

The following are a few sketches of relationships which are commonly used to model data. I have deliberately not included scales on the graphs because the scale will depend upon the application.

The relationships (functions) are defined in terms of parameters $a, A(>0), b, c, k(> 0)$ and n. Parameters are constants involved in the function equations whose values are to be determined from experimental results. It may be possible to give a mechanistic interpretation of these parameters (e.g. V_{max} in the M-M equation is the maximum reaction velocity) and this is desirable, though not always possible.

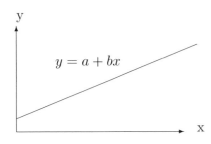

Straight line

Can be useful and often is.
But done to death because historically it was the only curve that could be fitted easily.
Use it when you know your data is linear or when you know nothing! It's a good start.

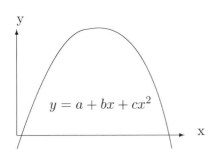

Polynomial

Mathematical.
Not very useful in terms of mechanistic modelling because interpretation of the parameters is difficult or impossible.
Easy to deal with mathematically.

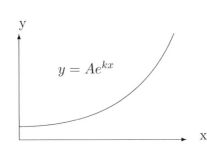

Exponential increase

The equation of growth.
Many applications in unrestricted growth.
e.g. bacteria on agar plate, epidemics, nuclear reaction.
Parameters are meaningful.

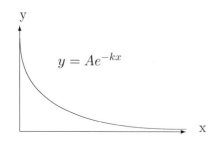

Exponential decay

The equation of radio-active decay.
Many applications in dispersion and diffusion
problems or modelling tracer experiments.
Parameters are meaningful.

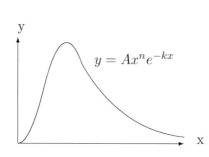

Logistic

Growth curve.
Population growth including limiting factors

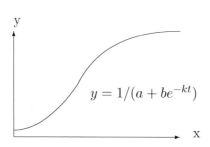

$Ax^n e^{-kx}$

Injection of pollutant or marker,
daily milk yield, etc.

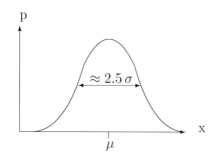

Normal frequency distribution

$$\frac{1}{\sqrt{2\pi\sigma^2}} \exp\left(\frac{(x-\mu)^2}{2\sigma^2}\right)$$

Chapter 5

Neat Tricks and Useful Solutions

I have chosen the topics in this chapter to introduce the more ambitious student to one or two more taxing but useful areas of mathematical thinking. They are intended to stimulate the mind and to give more of an insight into the way mathematicians approach problems.

It is not intended that you (the student) should read and memorise these examples. It is hoped that you will occasionally look at one or other of them, whether for reference - or fun - or inspiration.

5.1 The Difference of Two Squares

For reasons which are not obvious the expression $x^2 - a^2$ has great importance in school algebra. Students are asked to remember that

$$x^2 - a^2 = (x + a) \times (x - a)$$

When students ask (and they should ask) "Why does this relationship hold?" - they are lucky if they get a reply. And if they do it is most probably justified in the following way:
Because

$$
\begin{aligned}
(x + a)(x - a) &= x(x - a) + a(x - a) \\
&= x^2 - xa + ax - a^2 \\
&= x^2 - a^2
\end{aligned}
$$

Mathematically there is nothing wrong with the above, but

- it's the wrong way round

- it sheds no light on the nature of the problem.

47

Consider the following diagrams

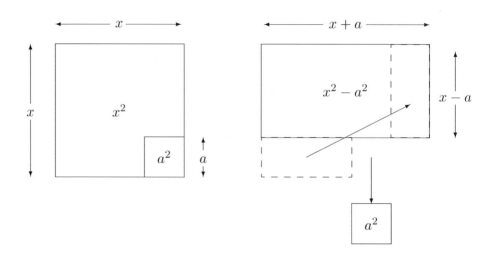

In the first diagram we have drawn two squares, one in the bottom right hand corner of the other. The right hand diagram shows the effect of removing the smaller square, leaving an object with area $x^2 - a^2$, and re-arranging what is left to make a rectangle with sides of length $x + a$ and $x - a$.

The moral of this story is that sometimes "a picture is worth a thousand words", and the equivalent mathematical viewpoint is that there are different ways of looking at a problem. In this case a simple bit of geometry can clarify a difficult bit of algebra.

5.2   Mathematical Induction

The discovery of scientific rules often results from an observation that individual results follow some sort of pattern. For example, the following relationships:-

$$\left.\begin{array}{rcccc} 1+3 & = & 4 & = & 2^2 \\ 1+3+5 & = & 9 & = & 3^2 \\ 1+3+5+7 & = & 16 & = & 4^2 \end{array}\right\} \qquad (5.1)$$

may lead to the conclusion that

$$1+3+5+\cdots+(2n-1) = n^2 \qquad (5.2)$$

where n is any +ve integer. Note that the observations in *equation* (5.1) only suggest that the general statement of *equation* (5.2)) may be true. How can we prove it? One method which can be used is known as mathematical induction and takes the following form.

1. show that the result is correct for a specific case ($n = 1$ say)

2. assume that the result is true for some arbitrary value ($n = k$ say) and show that the result is also true for $n = k + 1$.

3. combine the two previous results repeatedly to show that the formula is true for $n = 2, 3 \ldots$

Continuing the example mentioned above:-

Step 1 We have already observed that the result is correct for $n = 2, 3, 4 \ldots$ in *equation* (5.1). In the case where $n = 1$ the result is trivial.

Step 2 assume *equation* (5.2) is correct for $n = k$

$$\text{therefore} \sum_{i=1}^{k} (2i - 1) = k^2 \tag{5.3}$$

for n = k + 1

$$
\begin{aligned}
\sum_{i=1}^{k+1} (2i - 1) &= \sum_{i=1}^{k} (2i - 1) \quad \ldots \text{the first } k \text{ terms} \\
&\quad + (2k + 1) \qquad \ldots \text{the additional term} \\
&= k^2 + (2k + 1) \quad \text{from } \textit{equation } (5.3) \\
&= (k + 1)^2
\end{aligned}
$$

which is the same as the result of applying *equation* (5.3) with k replaced by $k + 1$. Thus if *equation* (5.2) holds for $n = k$, it must also hold for $n = k + 1$.

Step 3 *Equation* (5.2) must therefore hold for all +ve integers n since:

1. we know it to be true for $n = 1, 2, 3, 4$

2. if it is true for $n = 4$ it must be true for $n = 5$

3. successive applications of 'Step 2' for $n = 6, \ldots$ etc. will include all the positive integers.

5.3   Pythagoras' Theorem

You will all be familiar with Pythagoras' theorem, $a^2 = b^2 + c^2$, which has many applications in geometry. The hypotenuse is the longest side of a right-angled triangle, the one opposite the right-angle and the theorem is often stated as

> "The square on the hypotenuse is equal to the sum of the squares on the other two sides."

There are several proofs of the theorem, of which the following is probably the simplest.

Suppose we take a right-angled triangle and construct a square using four copies of it as in the diagram below.

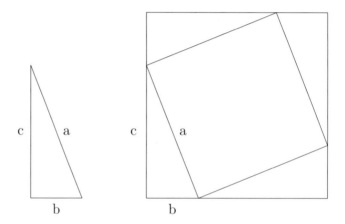

The area of the large square may be calculated using the lengths of its sides as:

$$Area = (b + c)^2$$

Its area may also be calculated as the area of the smaller square plus four times the area of the triangle abc:

$$Area = a^2 + 4 \times \left(\frac{b \times c}{2} \right) = a^2 + 2bc$$

Now we can equate the two expressions for the area

$$
\begin{aligned}
a^2 + 2bc &= (b + c)^2 \\
a^2 + 2bc &= b^2 + 2bc + c^2 \\
a^2 &= b^2 + c^2
\end{aligned}
$$

5.4 Pythagoras' Theorem revisited

The following proof, discovered by a young *Einstein* - who was supposedly backward, is particularly stimulating and elegant.

Suppose we take a triangle:

Any old triangle will do for starters. Then draw a "similar" (that means the same shape) triangle with sides twice as big:

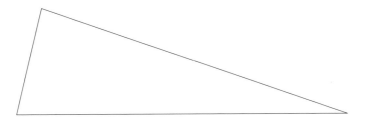

Is the area of the new triangle twice as big as the original, or if not how much bigger is it? In fact it's easy to show that it's actually four times as big:

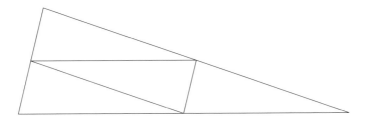

The original fits nicely into the new one four times exactly - the middle one is upside down but never mind that.

If we draw a new (similar) triangle with sides three times as big, it's easy to show that the area is nine times as big:

So, for a similar triangle with sides twice as big the area is 4 times as big, for a triangle with sides three times as big the area is 9 times as big. If the sides are four times as big, the area is 16 times as big, you might like to draw it to convince yourself.

From the preceding arguments it looks as though the area of similar triangles is proportional to the square of the lengths of their sides. We can represent this by a bit of mathematics:

$$A = ka^2$$

where A is the area, a is the length of one of the sides (the equivalent one) and k is some constant. The value of this constant will depend upon the shape of the triangle and upon which side (a) we are comparing. To evaluate it you must measure the area (A) of a specific triangle and its corresponding side length (a), then $k = A/a^2$.

Actually this rule applies to other shapes too, try proving it for rectangles - it's not so easy for other shapes but you can make most shapes out of a mixture of rectangles and triangles - if you try!

So, for a given shaped triangle we have

$$A = ka^2$$

Now consider the triangle below, which contains a right-angle at p. The line p-s is a perpendicular.

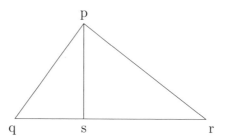

We have three similar triangles pqs, rqp and rps whose corresponding sides are a, b and c as shown below.

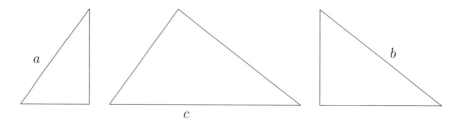

The areas will be ka^2, kc^2 and kb^2 respectively, where k is an unknown constant. In addition the area of the large triangle is made up from the areas of the two smaller ones. So that

$$kc^2 = ka^2 + kb^2$$

and therefore, dividing throughout by k we have

$$c^2 = a^2 + b^2$$

Pythagoras knew this, but his proof was much more difficult! The young Einstein, struggling with the original version, felt that there ought to be an easier way!

5.5   Limits

The concept of a mathematical limit will either strike you as something intuitively obvious or as something so obscure as to be unimportant. Unfortunately for those of you in the latter group limits are of fundamental importance in calculus so that a basic understanding will be beneficial.

Consider the function

$$y = \frac{1}{x}$$

As x takes larger and larger positive values, y becomes smaller and smaller. It is also obvious that however large we make x the value of y will never be negative. Thus we can say that as x increases, y decreases towards zero, and therefore that the limiting value of y as x tends to infinity will be zero. We write this as

$$\underset{x \to \infty}{Lim} \frac{1}{x} = 0$$

In general we shall be concerned with limits where x approaches some specific value. As an example

$$\underset{x \to 2}{Lim} (3 - x) = 1$$

This example is so obvious as to be almost meaningless (what is the point of it all!?). We can evaluate the limit simply by letting x take the limiting value and substituting in the function. Unfortunately however it is not always possible to do this. Consider:-

$$\underset{x \to 2}{Lim} \frac{x^2 - 2x}{x - 2}$$

If we substitute the value $x = 2$ into this function the resulting expression is

$$\frac{0}{0}$$

which is indeterminate [**Not**, in general, **Zero**]. However if we factorise the expression to give

$$\underset{x \to 2}{Lim} \frac{(x - 2)x}{(x - 2)} = x$$

we can see that, provided $x \neq 2$ we may cancel the factor $(x - 2)$ so that

$$\underset{x \to 2}{Lim} \frac{x^2 - 2x}{x - 2} = \underset{x \to 2}{Lim} x$$

and while x can <u>not</u> take the value of 2 exactly we may take x as close to the value 2 as we wish.

Thus we write:-

$$\underset{x \to 2}{Lim} \frac{x^2 - 2x}{x - 2} = 2$$

It is most important to understand that in evaluating the limiting value we have not simply calculated the function value at the limit - read this sentence again! In general we shall be able to evaluate the function as close to the limit as we choose but it may be impossible to calculate the value of the function at the limit. As close as we like - but not actually there. If this last paragraph isn't clear to you, read the section on limits again.

Example - The limiting value of $y = \frac{x^2 + x - 2}{x - 1}$ **as** $x \to 1$

$$\text{At } x = 1, \ y = \frac{1 + 1 - 2}{1 - 1} = \frac{0}{0}!!?$$

$$\text{Now } y = \frac{x^2 + x - 2}{x - 1} = \frac{(x - 1)(x + 2)}{x - 1}$$

$$\text{so that } \underset{x \to 1}{Lim} y = \underset{x \to 1}{Lim} \frac{(x - 1)(x + 2)}{x - 1}$$
$$= \underset{x \to 1}{Lim} (x + 2)$$
$$= 3$$

The concept of limiting values (limits) is the basis of calculus.

5.6 Trigonometry - angles with a difference

We use angles to describe the amount by which we rotate lines. For example we can rotate a line by a full revolution, in which case it returns to its original position, or we can rotate by a right-angle, in which case it will be perpendicular to its original position. We normally measure angles in units of *degrees*, in which a full rotation corresponds to 360 degrees. Thus a right-angle is equivalent to 90 degrees, or 90°. The following diagram contains an angle of approximately 45°.

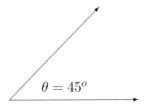

$$\theta = 45^o$$

In order to measure angles we use a protractor which is a semi-circular template with the angles marked around the curved side.

5.6.1   Radians and Degrees

Another method of measuring angles is to measure the length of the arc subtended by the angle and divide this by the radius. Using this method the angle is measured in units of *radians* . A full rotation (360 degrees) corresponds to 2π radians.

In science it is often more convenient to use radians than degrees so it is helpful to understand the difference. You calculator will work with either units, but if it is working in radians and you are measuring in degrees it will give the wrong answers, so watch out!

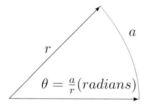

The following table shows equivalent angles in degrees and radians

degrees (^{o})	0	45	57.3	90	114.6	180	360
radians	0	$\pi/4$	1	$\pi/2$	2	π	2π

5.7   Trigonometric Ratios

The trigonometric functions can all be associated with ratios of the lengths of the various sides of a right-angled triangle. Consider the following right-angled triangle containing the angle θ.

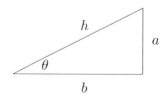

The values of the various trigonometrical functions (sine, cosine and tangent) are defined as follows:

$$\sin\theta = a/h = opposite/hypotenuse$$
$$\cos\theta = b/h = adjacent/hypotenuse$$
$$\tan\theta = a/b = opposite/adjacent$$

and are normally written in the above short forms.

Notice that as $\theta \to 0$, $\sin\theta \to \theta$ and $\cos\theta \to 1$, because $a \to h\theta$ and $b \to h$. See section 5.5 concerning limiting values.

Using the definitions above, together with Pythagoras' theorem, other useful relationships can be found such as:

$$\sin^2\theta + \cos^2\theta \;=\; 1$$

and
$$\tan\theta \;=\; \frac{\sin\theta}{\cos\theta}$$

In order to convince yourselves that you understand the trigonometric ratios prove the two relationships above for yourselves.

Radiation on a Surface

The intensity of radiation is usually measured in terms of the amount of radiation falling on an area perpendicular to the direction of radiation. Thus radiation of 100 W m^{-}2 implies that an area of one square meter at right-angles to the radiation would receive 100 W of radiative energy.

However, on a surface which is not at right-angles to the direction, for example the solar panel on your house roof, should you decide to go green:

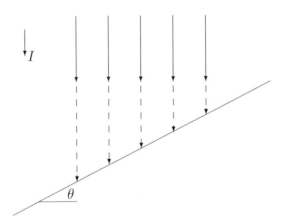

$$\text{intensity on inclined surface} \;=\; \text{Radiation Intensity} \times \frac{\text{area at right-angle}}{\text{area of shadow}}$$
$$=\; I \times \cos\theta$$

Chapter 6

Differential Calculus

6.1 Introduction

The word *calculus* is Latin for pebble, and in its generic form means any method of calculation. Therefore you have been making use of calculus for years! Early forms of the abacus used pebbles in grooves marked out in sand, hence the association.

However, in more recent times *the calculus* has come to mean the branch of mathematics which is concerned with the behaviour of dynamic systems, that is with systems in which objects move or change - like all living things: bacteria, cows and humans! It was developed by Fermat, Newton and others in order to study the motion of planets, pendulums... and falling apples?. Current applications include the modelling of plant/animal development and aspects of population growth, epidemiology etc. Most computer models are based on the methods of calculus, though they use numerical approximations in order to solve the (much) more complex equations necessary to describe these systems. The ability to construct differential equations that define such systems will allow you to make use of the many computer packages that can produce solutions.

Calculus comprises two processes; *differentiation* in which we know the equations defining the state of a system and use them to work out the rate at which the system will change as its independent variables change, and *integration* in which we know the equations defining the rate of change and we use them to predict the state of the system at specific values of its variables. Many people, especially on the island at the other side of the Atlantic, refer to integration as *anti-differentiation* - a hideous but usefully accurate term!

We begin by describing differential calculus, because differentiation can be defined using one formula - though working with it can be extremely tedious, whilst integration is more of an art form relying to a large extent on guesswork and experience gained from differentiation.

6.1.1   What is differentiation?

The following maths-speak introduces some complicated notation - don't panic! Just accept it for what it is - you will soon understand.

If we know a relationship $y = f(x)$, it is often possible to derive a formula that defines the slope of its graph at any point x. This formula, which we denote by dy/dx or $f'(x)$, is variously called the *derivative* or the *differential* or the *differential coefficient* of the function $y = f(x)$ and the process that we go through in order to find it is known (to mathematicians) as *differentiation*.

"Differentiation" is a word which causes me, and I suspect many of you, great difficulty, because its colloquial meaning gives little clue to its mathematical use. I am more at home with the word "differential", because I know that the differential on a car allows the right and left hand wheels to turn at different speeds when cornering. I am also familiar with the term "differential" referring to bicycles where the differential is the ratio of wheel velocity to pedal or crank velocity.

The differential on a bicycle is the result of dividing the number of teeth on the chainwheel (the one connected to the pedals), by the number of teeth on the rear wheel sprocket and tells you the relative speed (rpm) of the wheels corresponding to the speed (rpm) of the pedals. Thus if the chainwheel has 48 teeth and the rear wheel has 16 teeth, then the differential coefficient is 3, because for each turn of the pedals the wheels go round three times.

Also if we plot the angular speed of the wheels against the speed of the pedals we would see a straight line graph whose slope is 3. The differential coefficient is the slope of such a graph.

Thus the derivative of the function defining wheel velocity in terms of pedal velocity is the same as the differential, which is also the slope of the graph of wheel angular velocity against pedal velocity.

Differential coefficients are quantities or expressions that determine the relative change in a variable as its independent variable changes. Usually the independent variable will be time. Thus the growth rate is the change in mass, divided by the corresponding change in time. We would refer to this as dm/dt. Acceleration or the rate of change of velocity may also be expressed as a differential coefficient (dv/dt) i.e. the relative change in velocity as time changes. The independent variable will not necessarily always be time. For example we could ask what is the relative change in the area of a circle as the radius changes, in this case we would require an expression for dA/dr.

In general, if we can define a function which specifies the size, position, concentration, etc. of an object at a given time, then differentiation will enable us to derive equivalent functions for the growth rate, velocity, reaction rate, etc. of the object.

6.2 Distance and Velocity

Legend would have us believe that, one day whilst lying on his back in the orchard, Newton invented the formula:

$$s = \frac{1}{2}gt^2$$

which predicts the distance fallen s by a particle at time t from rest. (g is the gravitational constant which in SI units is 9.81m s^{-2}). Note that this equation ignores air resistance, but it was probably one of those sultry airless days!

We can use this equation to calculate s every quarter of a second during the first two seconds of its fall to give the following table:

t	0.00	0.25	0.50	0.75	1.00	1.25	1.50	1.75	2.00
s	0.00	0.31	1.23	2.76	4.91	7.66	11.04	15.02	19.62

from which we can plot a graph of s against t.

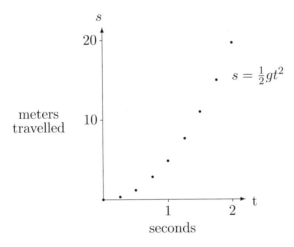

Figure 6.1: Distance against Time

We could fill in the gaps, either by calculating further values or by fitting a smooth curve through the existing points. Note that if we fit a curve through the points we are approximating, and that the alternative involves a *lot* of calculation!

Figure (6.1) allows us to see how far the particle has dropped at any given time, or it would if we filled in the gaps. However this in itself doesn't provide any more information than the original formula.

One question that we may like to answer is "What is the speed of the particle at any specified time?"

6.2.1   Average Velocity

The average velocity of the particle over a given time interval is given by the distance travelled divided by the time taken. This is an easy calculation. For example, if we want to know the average velocity over the first second $\bar{v}_{0,1}$, we can calculate it as follows:

$$\bar{v}_{0,1} = \frac{4.91 - 0.00}{1 - 0} = 4.91 \text{ m s}^{-1} \tag{6.1}$$

or if we wanted to calculate the average velocity between 1 and 2 seconds:

$$\bar{v}_{1,2} = \frac{19.62 - 4.91}{2 - 1} = 14.71 \text{ m s}^{-1} \tag{6.2}$$

You will observe that each answer is the slope of the chord joining the two points on the graph corresponding to the specified time interval.

6.2.2   Instantaneous Velocity

The instantaneous velocity is given by the slope of the tangent to the graph at a particular time, so we could calculate the velocity by drawing the tangent to the graph and then measuring its slope. This might be fun (?), but is certainly tedious if we are expected to produce accurate results or have to do it more than once. There must be an easier way!

Suppose we are required to find the velocity of the particle 1 second after it has been dropped.

One estimate of this velocity would be the average calculated over the interval 1 to 2 seconds. Observation says that a better estimate can be made using the average over the interval 1 to 1.5 seconds, and a better one still using the average over 1 to 1.25 seconds. *Table* (6.1) shows the results of using smaller and smaller intervals based on the time interval beginning at 1 second . An alternative approach is to calculate the average velocity over

time interval (t)	$\bar{v}$ (m s^{-1})
1.0000 to 2.0000	14.7150
1.0000 to 1.5000	12.2625
1.0000 to 1.2500	11.0363
1.0000 to 1.1000	10.3005
1.0000 to 1.0100	9.8590
1.0000 to 1.0010	9.8149
1.0000 to 1.0001	9.8105

Table 6.1: Average velocity over a small time period just after 1 second has elapsed

small time intervals just before 1 second as in *Table* (6.2).

time interval (t)	$\bar{v}$ (m s^{-1})
0.0000 to 1.0000	4.9050
0.5000 to 1.0000	7.3575
0.7500 to 1.0000	8.5838
0.9000 to 1.0000	9.3195
0.9900 to 1.0000	9.7610
0.9990 to 1.0000	9.8051
0.9999 to 1.0000	9.8096

Table 6.2: Average velocity over a small time interval just before 1 second elapses

It can be seen from these tables that the average velocity over the smaller intervals is converging to the value 9.81. More elaborate procedures using time limits straddling the specific time could be devised which converge more quickly, however the important thing to notice is that the values do converge to a limit, and that this limit is independent of the approach.

The numerical method used above is now a standard method for evaluating *rates*, but in Newton's time, and for a considerable period afterwards (until the 1930's) the arithmetic was too time consuming.

Newton and his contemporaries didn't have the advantage of modern computers to solve such problems so they looked for more convenient analytical methods. This was where "the calculus" was conceived. The calculus solution follows a similar process to the above but avoids specific numbers in order to find a general formula for the velocity at any value of t.

Earlier we estimated the instantaneous velocity at time t by calculating the average velocity over a small time interval beginning at t. The mathematicians way of saying this is to look at the time interval between t and $t+\delta t$ where t is any value of time and δt is a small interval. (Mathematicians use the Greek letter δ (delta) as shorthand for "a small amount of . . .".)

In order to calculate the average velocity we need to calculate the distance travelled (δs) during the time interval δt.

$$\delta s = \frac{1}{2}g(t + \delta t)^2 - \frac{1}{2}gt^2 \qquad (6.3)$$

The estimate of velocity at time t is given by

$$
\begin{aligned}
v(t) = \frac{\delta s}{\delta t} &= \frac{\frac{1}{2}g(t + \delta t)^2 - \frac{1}{2}gt^2}{\delta t} \\
&= \frac{g\{(t + \delta t)^2 - t^2\}}{2\delta t} \\
&= \frac{g\{t^2 + 2t\delta t + \delta t^2 - t^2\}}{2\delta t} \\
&= \frac{g\{2t\delta t + \delta t^2\}}{2\delta t} \\
&= gt + \frac{1}{2}g\delta t
\end{aligned}
$$

This is a general expression for *equations* 6.1 *and* 6.2, giving the average velocity between t and $t + \delta t$ seconds.

Now we know that the best estimate of the instantaneous velocity is calculated when the time interval is made as small as possible. The mathematician's way of saying this is when $\delta t \to 0$ (*ie* the time interval approaches zero).

If we let $\delta t \to 0$ the second term in the equation disappears (because we can make it small enough to ignore) and we are left with

$$
\text{velocity} = \underset{\delta t \to 0}{Lim} \frac{\delta s}{\delta t} = gt
$$

As usual, mathematicians have a short-hand way of writing $\underset{\delta t \to 0}{Lim} \frac{\delta s}{\delta t}$, they write this as $\frac{ds}{dt}$ and call it the *differential coefficient* of s with respect to t, so we have

$$
\text{velocity} = \frac{ds}{dt} = gt \tag{6.4}
$$

It can be seen that when $t = 1$, $\frac{ds}{dt} = 9.81$, which agrees with our previous calculation. The value of the velocity at $t = 0$ is also easy to calculate ($= 0$) and agrees with intuition and our graph (*Figure* (6.1)). Thus we have some confidence in *equation* (6.4) and are now in a position to be able to calculate the velocity at any instant.

[**Pause for meditation:** If the differential of $gt^2/2$ is gt then the anti-differential of gt must be something to do with $gt^2/2$?]

6.3   The Differential Coefficient of any function

The calculation of velocity carried out in the previous section can be generalised to provide an expression for the differential coefficient of any function.

Suppose we are given the relationship $y = f(x)$ where $f(x)$ is some specified function of x.

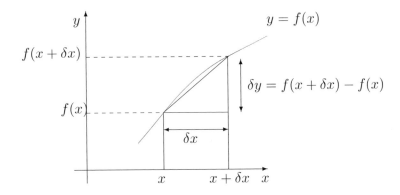

Figure 6.2: Estimating the slope of $y = f(x)$ at $(x, f(x))$

In this case we look at two values of the function at x and $x + \delta x$. We can calculate the corresponding y values as $f(x)$ and $f(x+\delta x)$ and use them to calculate the change δy in y corresponding to the change δx in x.

The slope of the curve $y = f(x)$ at the point (x, y) is now given by

$$\frac{dy}{dx} = \underset{\delta x \to 0}{\text{Lim}} \frac{\delta y}{\delta x} = \underset{\delta x \to 0}{\text{Lim}} \frac{f(x + \delta x) - f(x)}{\delta x} \tag{6.5}$$

Equation (6.5) defines the *differential coefficient* of y with respect to x, and is the basis of differential calculus - all of what follows is dependent upon it. Working with this equation can be extremely tedious and time consuming as you will see from some of the following examples. However, tables of standard derivatives exist and there are methods and short cuts which make life easier. We shall develop some of these in the following sections. If you can though, you should try to understand what differentiation is about at least once: there is no substitute for knowing what is going on.

Example - The differential coefficient of the function $y = 3x$

$$
\begin{aligned}
\frac{dy}{dx} &= \lim_{\delta x \to 0} \frac{3 \times (x + \delta x) - 3 \times (x)}{\delta x} \\
&= \lim_{\delta x \to 0} \frac{3x + 3\delta x - 3x}{\delta x} \\
&= \lim_{\delta x \to 0} \frac{3\delta x}{\delta x} \\
&= 3
\end{aligned}
$$

This answer tells us that the slope of the function $y = 3x$ is 3, irrespective of the value of x, *ie* it is a straight line. This should come as no surprise to you, but if you understand it you can give yourself a tick. The maths looks horrid, but the ideas aren't too bad if you just dig your way through the jargon. In general you can ignore the jargon, but sometimes it is helpful to be precise.

Example - The derivative of the function $y = x^2 + 3x + 2$

$$
\begin{aligned}
\frac{dy}{dx} &= \lim_{\delta x \to 0} \frac{[(x + \delta x)^2 + 3(x + \delta x) + 2] - [x^2 + 3x + 2]}{\delta x} \\
&= \lim_{\delta x \to 0} \frac{x^2 + 2x\delta x + \delta x^2 + 3x + 3\delta x + 2 - x^2 - 3x - 2}{\delta x} \\
&= \lim_{\delta x \to 0} \frac{(2x + \delta x + 3)\delta x}{\delta x} \\
&= \lim_{\delta x \to 0} 2x + 3 + \delta x \\
&= 2x + 3
\end{aligned}
$$

Example - the derivative of $\sqrt{x}$.

$$
\frac{d}{dx}\sqrt{x} = \frac{d}{dx}x^{\frac{1}{2}}
$$

In this case it is convenient to express the estimate of slope in a different way. An alternative form of the equation is

$$
\frac{\delta y}{\delta x} = \frac{f(q) - f(x)}{q - x} \qquad \text{think of } q \text{ as } x + \delta x
$$

so that we have

$$\frac{\delta y}{\delta x} = \frac{q^{\frac{1}{2}} - x^{\frac{1}{2}}}{q - x}$$

$$= \frac{q^{\frac{1}{2}} - x^{\frac{1}{2}}}{(q^{\frac{1}{2}} - x^{\frac{1}{2}})(q^{\frac{1}{2}} + x^{\frac{1}{2}})} \quad [since\ a^2 - b^2 = (a - b)(a + b)]$$

$$\therefore \frac{d}{dx} x^{\frac{1}{2}} = \lim_{q \to x} \frac{1}{q^{\frac{1}{2}} + x^{\frac{1}{2}}}$$

$$= \frac{1}{2x^{\frac{1}{2}}} = \frac{1}{2} x^{-\frac{1}{2}}$$

$$\therefore \frac{d}{dx} x^{\frac{1}{2}} = \frac{1}{2} x^{-\frac{1}{2}}$$

6.3.1 Differentiability

A function $y(x)$ is said to be differentiable if the change δy in y, corresponding to a change δx in x can be made arbitrarily small by choosing δx as small as we wish.

Most functions which describe physical situations are differentiable. Such functions are often referred to as being "well behaved", "smooth", "continuous" etc.

The following is an example of a function which is not differentiable:-

$$x = \begin{cases} 1 & \text{if } x > 0 \\ -1 & \text{otherwise} \end{cases}$$

This function is not differentiable at $x = 0$ and is thus called discontinuous.

The physical significance of the above is that a function will be differentiable provided it has no sharp corners or values which are infinitely large.

6.4   Differentials involving two Functions

The evaluation of complicated derivatives can be extremely tedious and difficult, especially if we have to resort to first principles using *equation* (6.5). It will be useful to develop a few shortcuts and tools which may ease the process. The following is a formal treatment showing how the rules for differentiating the sum, product, and quotient of two functions are developed. Whilst the rules are important, their derivation may be found tedious and may be skipped over by the faint hearted. The applications and examples should be mastered however.

Consider two differentiable functions of x: $u = u(x)$ and $v = v(x)$.

$$
\begin{aligned}
\text{Let} \qquad u + \delta u &= u(x + \delta x) \\
\text{and} \qquad v + \delta v &= v(x + \delta x)
\end{aligned}
$$

so that δu is the increase in the value of the function u as x increases by the amount δx, and δv is the corresponding increase in v.

6.4.1   The derivative of a sum $u(x) + v(x)$

$$
\begin{aligned}
\frac{d}{dx}(u + v) &= \underset{\delta x \to 0}{Lim} \frac{[u(x + \delta x) + v(x + \delta x)] - [u(x) + v(x)]}{\delta x} \\
&= \underset{\delta x \to 0}{Lim} \frac{u + \delta u + v + \delta v - u - v}{\delta x} \\
&= \underset{\delta x \to 0}{Lim} \frac{u + \delta u - u + v + \delta v - v}{\delta x} \\
&= \underset{\delta x \to 0}{Lim} \frac{\delta u + \delta v}{\delta x} \\
&= \underset{\delta x \to 0}{Lim} \frac{\delta u}{\delta x} + \underset{\delta x \to 0}{Lim} \frac{\delta v}{\delta x}
\end{aligned}
$$

Thus

$$
\frac{d}{dx}(u + v) = \frac{du}{dx} + \frac{dv}{dx} \tag{6.6}
$$

Example - the derivative of $x^{1/2} + 3x$

$$
\frac{d}{dx}(x^{1/2} + 3x) = \frac{d}{dx}x^{1/2} + \frac{d}{dx}3x = \frac{1}{2}x^{-1/2} + 3
$$

6.4.2 The derivative of a product $u(x)v(x)$

$$\frac{d}{dx}(uv) = \underset{\delta x \to 0}{Lim} \frac{(u + \delta u)(v + \delta v) - uv}{\delta x}$$

$$= \underset{\delta x \to 0}{Lim} \frac{uv + u\delta v + v\delta u + \delta u\delta v - uv}{\delta x}$$

$$= \underset{\delta x \to 0}{Lim} u\frac{\delta v}{\delta x} + \underset{\delta x \to 0}{Lim} v\frac{\delta u}{\delta x} + \underset{\delta x \to 0}{Lim} \frac{\delta u\delta v}{\delta x}$$

$$= u\frac{dv}{dx} + v\frac{du}{dx} + \underset{\delta x \to 0}{Lim} \frac{\delta u\delta v}{\delta x}$$

The final term may be treated as either

$$\underset{\delta x \to 0}{Lim} \delta u\frac{\delta v}{\delta x} = \delta u\frac{dv}{dx}$$

or

$$\underset{\delta x \to 0}{Lim} \delta v\frac{\delta u}{\delta x} = \delta v\frac{du}{dx}$$

In either case it will be zero since as $\delta x \to 0$ both $\delta u \to 0$ and $\delta v \to 0$ while

$$\frac{dv}{dx} \ and \ \frac{du}{dx}$$

will take limiting values so that:

$$\frac{d}{dx}(uv) = u\frac{dv}{dx} + v\frac{du}{dx} \tag{6.7}$$

Example - The derivative of x^2

If we let both $u = x$ and $v = x$ we can use *equation* (6.7) to calculate the derivative of x^2 as follows:

$$\frac{d}{dx}(x.x) = x.1 + 1.x = 2x$$

since the slopes of both $u = x$ and $v = x$ are 1, and hence there derivatives are 1. This result tells us that the slope of the curve $y = x^2$ has the value $2x$. Thus when $x = 3$, $y = 9$ and the slope of the curve is 6.

Example - The derivative of x^3

The same process can be used here:

$$\frac{d}{dx}(x^2.x) = x^2.1 + 2x.x = 3x^2$$

It is left as a challenge to the student to prove that $\frac{d}{dx}x^n = nx^{n-1}$ using the method of induction. See section 5.2.

6.4.3   The derivative of a quotient $u(x)/v(x)$

$$\frac{d}{dx}\left(\frac{u}{v}\right) = \underset{\delta x \to 0}{Lim} \frac{\frac{u+\delta u}{v+\delta v} - \frac{u}{v}}{\delta x}$$

$$= \underset{\delta x \to 0}{Lim} \frac{uv + v\delta u - uv - u\delta v}{(v + \delta v)v\delta x}$$

$$= \underset{\delta x \to 0}{Lim} \frac{v\delta u - u\delta v}{(v + \delta v)v\delta x}$$

$$= \underset{\delta x \to 0}{Lim} \frac{v\frac{\delta u}{\delta x} - u\frac{\delta v}{\delta x}}{(v + \delta v)v}$$

$$\frac{d}{dx}\left(\frac{u}{v}\right) = \frac{v\frac{du}{dx} - u\frac{dv}{dx}}{v^2} \qquad (6.8)$$

6.5   Some important derivatives

The calculation of derivatives of complicated expressions will be eased if we use the rules developed in the previous section in conjunction with a list of standard derivatives. Tables of derivatives are available, but much can be done with a small selection of derivatives and we shall show the evaluation of the most important of these in this section. The derivation of these differential coefficients is of no great importance in itself, though some of you may like to know how the mathematical arguments go. However, it will be useful to follow the arguments through at least once, if only to gain insight into the methods of mathematical proof.

6.5.1   The derivative of a Constant

If $y = A$:

$$\frac{d}{dx}A = \underset{\delta x \to 0}{Lim} \frac{f(x + \delta x) - f(x)}{\delta x}$$

$$= \underset{\delta x \to 0}{Lim} \frac{A - A}{\delta x} = 0$$

$$\frac{d}{dx}\text{constant} = 0$$

This result is obvious since the slope (and hence the derivative) of the line $y = constant$ is zero.

6.5.2 The derivative of x^n

$$\frac{d}{dx}x^n = \underset{\delta x \to 0}{Lim}\frac{(x+\delta x)^n - x^n}{\delta x}$$

$$(x + \delta x)^n = x^n + nx^{n-1}\delta x + \frac{n(n-1)}{2!}x^{n-2}\delta x^2 + \cdots + nx\delta x^{n-1} + \delta x^n$$
$$= x^n + nx^{n-1}\delta x + 0(\delta x^2)$$

The expression $0(\delta x^2)$ is a shorthand way of saying "all the remaining terms which involve powers of δx^2 or greater".

$$\frac{d}{dx}x^n = \underset{\delta x \to 0}{Lim}\frac{x^n + nx^{n-1}\delta x + 0(\delta x^2) - x^n}{\delta x}$$
$$= \underset{\delta x \to 0}{Lim}nx^{n-1} + 0(\delta x)$$

[All the terms which involved δx^2 will now involve δx since we divide them all by δx]

$$= nx^{n-1}$$

since all the terms involving δx will tend to zero.

$$\text{Thus } \frac{d}{dx}x^n = nx^{n-1}$$

6.5.3 The derivative of $\sin x$

$$\frac{d}{dx}\sin x = \underset{\delta x \to 0}{Lim}\frac{\sin(x+\delta x) - \sin x}{\delta x}$$
$$= \underset{\delta x \to 0}{Lim}\frac{\sin x \cos \delta x + \cos x \sin \delta x - \sin x}{\delta x} \quad (6.9)$$

because, by reference to *equation* (8.5) in the chapter on matrix algebra later:

$$\sin(\theta + \phi) = \sin\theta\cos\phi + \cos\theta\sin\phi$$

Consider the following diagram:-

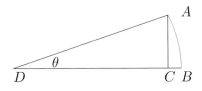

If θ is measured in radians

$$\theta = \frac{AB}{BD} = \frac{\text{length of arc}}{\text{radius}}$$

$$sin\theta = \frac{AC}{BD}$$

From inspection: as $\theta \to 0$ $AC \to AB$

$$\therefore \ \underset{\theta \to 0}{Lim} \ sin\,\theta = \theta$$

$$cos\theta = \frac{DC}{AD} = \frac{DC}{DB}$$

From inspection: as $\theta \to 0$ $DC \to DB$

$$\therefore \ \underset{\theta \to 0}{Lim} \ cos\,\theta = 1$$

Using these two limits we can rewrite *equation* (6.9)

$$\frac{d}{dx}\,sin\,x \;=\; \underset{\delta x \to 0}{Lim}\,\frac{sin\,x.1 + cos\,x\delta x - sin\,x}{\delta x}$$

$$= \; \underset{\delta x \to 0}{Lim}\,cos\,x$$

$$= \; cos\,x$$

$$\frac{d}{dx}\,sinx = cosx$$

By similar arguments we can find the derivative of $cos\,x$

$$\frac{d}{dx}\,cosx = -sinx$$

6.5.4   The derivative of a constant times a function of x

$$\frac{d}{dx}\,af(x) \;=\; \underset{\delta x \to 0}{Lim}\,\frac{af(x + \delta x) - af(x)}{\delta x}$$

$$= \; \underset{\delta x \to 0}{Lim}\,\frac{a[f(x + \delta x) - f(x)]}{\delta x}$$

$$= \; a\,\underset{\delta x \to 0}{Lim}\,\frac{f(x + \delta x) - f(x)}{\delta x}$$

$$= \; a\frac{d}{dx}f(x)$$

$$\therefore \ \frac{d}{dx}\,af(x) = a\frac{d}{dx}\,f(x)$$

Example - The derivative of $3\sin x$

$$= 3\frac{d}{dx}\sin x = 3\cos x$$

Example - The derivative of $4x^2$

$$\frac{d}{dx}4x^2 = 4\frac{d}{dx}x^2$$
$$= 4 \times 2x = 8x$$

[**Thinks!** if the derivative of $4x^2$ is $8x$, then so is the derivative of $4x^2 + c$ (where c is an unknown constant). So the integral (anti-differential) of $8x$ may be $4x^2 + c$?]

Example - The derivative of $x\sin x$

Let $x\sin x = uv$ where $u = x$ and $v = \sin x$

$$\frac{d}{dx}uv = u\frac{dv}{dx} + v\frac{du}{dx} \qquad \text{using } equation \text{ (6.7)}$$

$$\therefore \frac{d}{dx}x\sin x = x\cos x + \sin x \times 1$$
$$= \sin x + x\cos x$$

Example - The derivative of $\tan x$

$$y = tanx = \frac{\sin x}{\cos x} = \frac{u}{v}$$

$$u = \sin x \qquad \frac{du}{dx} = \cos x$$

$$v = \cos x \qquad \frac{dv}{dx} = -\sin x$$

$$\frac{d}{dx}\left(\frac{u}{v}\right) = \frac{v\frac{du}{dx} - u\frac{dv}{dx}}{v^2} \qquad \text{using } equation \text{ (6.8)}$$

$$\therefore \frac{d}{dx}\tan x = \frac{\cos x . \cos x - \sin x(-sinx)}{\cos^2 x}$$

but $\qquad \sin^2 x + \cos^2 x = 1 \qquad$ from *section* (5.6)

$$\therefore \frac{d}{dx}\tan x = \frac{1}{\cos^2 x}$$

6.5.5   The derivative of e^x

Consider the function a^x where a is an arbitrary constant.

$$
\begin{aligned}
\frac{d}{dx}\, a^x &= \underset{\delta x \to 0}{Lim}\ \frac{a^{x+\delta x} - a^x}{\delta x} \\[2mm]
&= \underset{\delta x \to 0}{Lim}\ a^x \frac{(a^{\delta x} - 1)}{\delta x} \\[2mm]
&= a^x\ \times\ \underset{\delta x \to 0}{Lim}\ \frac{a^{\delta x} - 1}{\delta x}
\end{aligned}
$$

The limit is difficult to evaluate but we can show that it exists if we look at a few explicit examples.

$$
\text{consider } \frac{d}{dx}\, 2^x = 2^x\ \times\ \underset{\delta x \to 0}{Lim}\ \frac{2^{\delta x} - 1}{\delta x}
$$

δx	$\frac{2^{\delta x}-1}{\delta x}$
0.1	0.717735
0.01	0.695555
0.001	0.693387
0.0001	0.693169

It can be seen that as $\delta x \to 0$ the value of the expression converges to 0.693147.

$$
\text{Similarly } \frac{d}{dx}\, 2.5^x\ \approx\ 2.5^x\ \times\ 0.916291
$$

$$
\text{and } \frac{d}{dx}\, 3^x\ \approx\ 3^x\ \times\ 1.098612
$$

It appears that there should be some value of a such that $\frac{d}{dx} a^x = a^x$ and that this value must lie somewhere between 2.5 and 3.0. In fact this number turns out to be the value $2.718282\ldots$ which is the exponential constant (e) and is sometimes referred to as Euler's number after the scientist who discovered it .

$$
\text{Thus } \frac{d}{dx}\, e^x = e^x \tag{6.10}
$$

The exponential function $\exp(x) \equiv e^x$ may be expressed in another form.

$$
\begin{aligned}
\exp(x) &= 1 + x + \frac{x^2}{2!} + \frac{x^3}{3!} + \cdots + \frac{x^n}{n!} + \cdots \\[2mm]
&= \sum_{i=0}^{\infty} \frac{x^i}{i!}
\end{aligned}
$$

where $n!$ (referred to as "factorial n") is calculated as $n(n-1)(n-2)\cdots(2)(1)$, for example $3! = (3)(2)(1) = 6$.

The derivative of $\exp(x)$ is therefore the sum of the derivatives of the individual terms in the expansion.

$$\exp(x) = 1 + x + \frac{x^2}{2!} + \frac{x^3}{3!} + \cdots$$

$$\frac{d}{dx}\exp(x) = 0 + 1 + \frac{2x}{2!} + \frac{3x^2}{3!} + \cdots$$

$$= 0 + 1 + x + \frac{x^2}{2!} + \cdots$$

$$= \exp(x) \qquad \text{since the R.H.S.} = \sum_{i=0}^{\infty}\frac{x^i}{i!}$$

6.5.6   The derivatives of $\ln x$ and a^x

$$\text{If } y = \ln x$$

$$x = e^y$$

$$\text{so that } \frac{dx}{dy} = e^y$$

$$\text{but } \frac{dy}{dx}\frac{dx}{dy} = 1$$

$$\therefore \frac{dy}{dx} = \frac{1}{\left(\frac{dx}{dy}\right)} = \frac{1}{e^y} = \frac{1}{x}$$

$$\frac{d}{dx}\ln x = \frac{1}{x}$$

We have seen that the derivative of a^x (where a is a constant) is difficult to evaluate from first principles. The problem is relatively simple now, however, since we can use logarithms. The derivative is found as follows:-

$$\text{if } y = a^x$$

$$\ln y = x\ln a \qquad\qquad (\ln \equiv log_e)$$

$$\therefore x = \frac{\ln y}{\ln a}$$

$$\text{and } \frac{dx}{dy} = \frac{1}{\ln a}\left(\frac{d}{dx}\ln y\right) = \frac{1}{y\ln a}$$

$$\therefore \frac{dy}{dx} = y\ln a$$

$$\frac{d}{dx}(a^x) = a^x \ln a$$

6.5.7 The Chain Rule

So far we have been limited in the types of function which we can differentiate. For example we can differentiate the function $y = \sin x + x^2$ but we are unable to differentiate $y = \sin(x^2)$. The latter is an example of a function of a function.

Suppose that u is a function of x

$$u = f(x)$$

and that y is a function of u

$$y = g(u)$$

so that we may write

$$y = g(f(x))$$

Using the example mentioned already we would have

$$u = x^2$$

$$y = \sin(u) = \sin(x^2)$$

Now if we let x increase by a small amount to $x + \delta x$, we cause u to change from u to $u + \delta u$ and therefore y will change from y to $y + \delta y$. We may write the following algebraic equation relating these small increments:

$$\frac{\delta y}{\delta x} = \frac{\delta y}{\delta u}\frac{\delta u}{\delta x}$$

Now if we let $\delta x \to 0$ this equation becomes

$$\frac{dy}{dx} = \frac{dy}{du}\frac{du}{dx}$$

Notice that by cancelling the two du terms, the above equation is obvious!

Example - The derivative of $\sin(x^2)$

$$
\begin{aligned}
y &= \sin(x^2) \\
\text{put } u &= x^2 \\
\text{so that } y &= \sin u \\
\text{therefore } \frac{dy}{dx} &= \frac{dy}{du}\frac{du}{dx} = \cos u \times 2x \\
\frac{d}{dx}\sin x^2 &= 2x \cos x^2
\end{aligned}
$$

I sometimes find it easier to write the function explicitly, rather than referring to an additional (abstract) function as follows:

$$\frac{d}{dx}\sin x^2 = \frac{d\sin x^2}{dx^2}\cdot\frac{dx^2}{dx}$$
$$= \cos x^2.2x$$

This is exactly the same as the above - but I prefer to leave out the "put $u = x^2$ etc." since it makes the solution longer and messy!

Example - The derivative of $(x^2 + 3x + 1)^4$

$$\begin{aligned} y &= (x^2 + 3x + 1)^4 \\ \text{put } u &= x^2 + 3x + 1 & \frac{du}{dx} &= 2x + 3 \\ y &= u^4 & \frac{dy}{du} &= 4u^3 \\ \frac{dy}{dx} &= \frac{dy}{du}\frac{du}{dx} \\ &= 4u^3(2x + 3) \\ \frac{dy}{dx} &= 4(x^2 + 3x + 1)^3(2x + 3) \end{aligned}$$

As in the previous example, the solution is shorter if we write it without using the abstract function:

$$\frac{d}{dx}(x^2 + 3x + 1)^4 = \frac{d(x^2 + 3x + 1)^4}{d(x^2 + 3x + 1)}\cdot\frac{d(x^2 + 3x + 1)}{dx}$$
$$= 4(x^2 + 3x + 1)^3.(2x + 3)$$

Example - the derivative of e^{kt}

$$\frac{d}{dt}e^{kt} = \frac{de^{kt}}{d(kt)}\cdot\frac{d(kt)}{dt}$$
$$= e^{kt}.k$$
$$\frac{d}{dt}e^{kt} = ke^{kt} \tag{6.11}$$

This is an important result with applications in many areas of biology, particularly in describing population growth and rates of decay of pollutants, trace elements, etc. It states that the rate of increase/decrease of an exponential function is proportional to the function itself. i.e. the more you have the bigger the rate of increase/decrease!

6.6 Optimum values - maxima and minima

We are often requested to find the optimum (maximum or minimum) value of some function or process. For example the manager is much more interested in the maximum profit available than he is in any other value of profit. We might also be asked to define the minimum cost of production or the maximum growth rate etc.

Consider the function plotted below:-

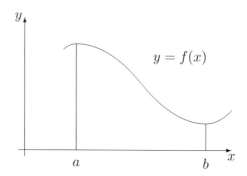

The values of y at $x = a$ and $x = b$ are maximum and minimum respectively because all other values in the vicinity of $x = a$ are less than $f(a)$, and all the values in the region of $x = b$ are greater than $f(b)$. Note that the value of the function at $x = a$ and $x = b$ are not necessarily the largest or smallest values that the function can take. A function may have any number of maxima and minima.

It is perhaps stating the obvious that maxima and minima occur alternatively for any function. After a maximum value the function must decrease before increasing again to the next maximum, therefore there must be some point between two maxima where a minimum value occurs. Similarly there must be a maximum value between each pair of minima.

Consider the graphs in *figure* (6.3) where the function and its first and second derivatives are plotted. The second derivative $\frac{d^2y}{dx^2} = \frac{d}{dx}\frac{dy}{dx}$ is the slope of $\frac{dy}{dx}$ or the "slope of the slope". Note the values of these derivatives given in the following table.

	$\frac{dy}{dx}$	$\frac{d^2y}{dx^2}$
maximum	0	$-ve$
minimum	0	$+ve$
inflexion	0	0

We can use the above to identify the turning points of any specific function. Non-zero values of the second derivative allow us to identify maxima or minima but a zero value does not necessarily identify an inflexion. In order to be certain the higher order derivatives must be evaluated until a non zero value is found. If $\frac{d^n y}{dx^n}$ is the first non-zero derivative (other than $\frac{dy}{dx}$) then

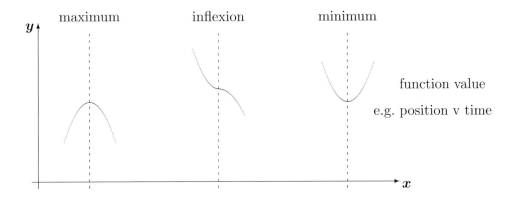

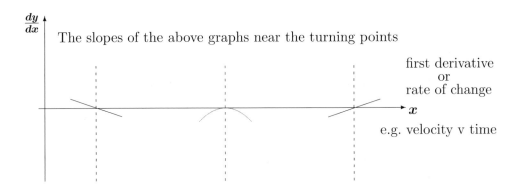

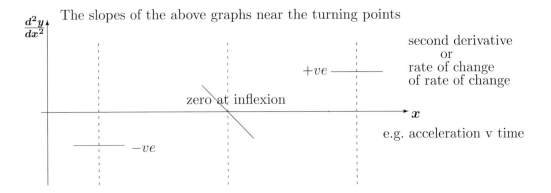

Figure 6.3: Turning points and their first two derivatives

the turning point is an inflexion if n is odd. Otherwise it will be a maximum or minimum depending upon whether $\frac{d^n y}{dx^n}$ is negative or positive.

Example - How fast should a fish swim?

If the rate of energy utilisation of a fish swimming is proportional to the cube of its speed, show that the most economical speed for the fish to swim against the current will be $1\frac{1}{2}$ times the current.

Let speed of the current be R and the speed of the fish relative to the water be S. The speed of the fish relative to the bank B is given by

$$B = S - R$$

Over a given distance x the time taken will be

$$t = \frac{x}{B} = \frac{x}{S - R}$$

and since the rate of energy utilisation is kS^3, where k is a constant, the total energy used (E) in covering the distance x is given by

$$E = kS^3 \frac{x}{S - R} = kx \frac{S^3}{S - R}$$

The minimum value of E is obtained when $\frac{dE}{dS} = 0$

$$\text{i.e.} \quad \frac{d}{dS} \frac{S^3}{S - R} = 0$$

$$\frac{(S - R)3S^2 - S^3}{(S - R)^2} = 0$$

$$\therefore \quad 3(S - R) - S = 0$$

$$2S = 3R$$

$$\therefore \quad S = \frac{3R}{2}$$

6.7 Small Errors

From the definition of the differential coefficient

$$\frac{dy}{dx} = \underset{\delta x \to 0}{Lim} \frac{\delta y}{\delta x}$$

we can see that provided δx is small

$$\frac{\delta y}{\delta x} \approx \frac{dy}{dx}$$

and so by multiplying both sides by δx we have:

$$\delta y \approx \frac{dy}{dx} \delta x$$

Inevitably there are errors in measurement during experimental processes, and the use of the differential coefficient as above allows us to calculate the error on derived variables. This can be an important part of the analysis when comparing errors of measurement with those associated with different treatments. See the calculation below.

Example - Estimation of Errors

In a water droplet experiment we need to calculate the volume of the droplet using a microscope to measure the diameter $D(mm)$. If we can measure the diameter to an accuracy of $\pm$ 0.001 mm what is the accuracy of our calculated volume, assuming that the droplets are spherical?

$$
\begin{aligned}
V &= \frac{4}{3}\pi(\frac{D}{2})^3 \\
&= \frac{\pi}{6}D^3 \\
\frac{dV}{dD} &= \frac{3\pi D^2}{6} = \frac{\pi D^2}{2} \\
\therefore \quad \frac{\delta V}{\delta D} &\approx \frac{\pi D^2}{2} \\
\delta V &\approx \frac{\pi D^2}{2}\delta D \\
&\approx \pm 0.0005\pi D^2 mm^3
\end{aligned}
$$

6.8   Summary Notes on Differentiation

6.8.1   Standard Derivatives

$$\frac{d}{dx} f(x) = \underset{\delta x \to 0}{Lim} \frac{f(x + \delta x) - f(x)}{\delta x}$$

$$\frac{d}{dx} \ln x = \frac{1}{x}$$

$$\frac{d}{dx} \text{constant} = 0$$

$$\frac{d}{dx} x^n = n x^{n-1}$$

$$\frac{d}{dx} \sin x = \cos x$$

$$\frac{d}{dx} \cos x = -\sin x$$

$$\frac{d}{dx} e^x = e^x$$

6.8.2   Rules for Differentiation

If u and v are both functions of x:-

$$\frac{d}{dx}(u + v) = \frac{du}{dx} + \frac{dv}{dx}$$

$$\frac{d}{dx}(u.v) = u\frac{dv}{dx} + v\frac{du}{dx}$$

$$\frac{d}{dx}\left(\frac{u}{v}\right) = \frac{v\frac{du}{dx} - u\frac{dv}{dx}}{v^2}$$

If y is a function of θ and θ is a function of x:-

$$\frac{dy}{dx} = \frac{dy}{d\theta}\frac{d\theta}{dx}$$

6.8.3   Maxima and Minima

At a **maximum** or **minimum** value of y, $\frac{dy}{dx} = 0$.

At a **maximum** $\frac{d^2y}{dx^2}$ is **negative**, at a **minimum** it is **positive**.

More generally, if $\frac{dy}{dx} = 0$ and the first non-zero derivative is odd ($\frac{d^3y}{dx^3}$, $\frac{d^5y}{dx^5}$, etc.) there is an inflexion. However, if it is even the turning point is a maximum or a minimum depending upon whether the sign of this derivative is negative or positive.

6.9 Applications

Equation for Radio-active Decay

Radio-active substances are used extensively to trace various chemical and biological reactions in living organisms. Accidents such as those at Chernobyl and Windscale (Sellafield) also provide ample opportunity to examine the effects of radio-active pollution. Here we derive the fundamental equation for exponential decay.

Let $R(t)$ Bq be the radiation at time t and R_0 Bq the initial radiation.[1]

The rate of increase $(\frac{dR}{dt})$ of radio-active material will be negative, and proportional to the amount of radio-active substance present :-

$$\frac{dR}{dt} \propto -R \text{ or } \frac{dR}{dt} = kR$$

where k is a negative constant, the *decay constant*, specific to the material.

From our experience, or reference to the earlier example (*equation* (6.11)), we can remember that

$$\frac{d}{dt} e^{kt} = k e^{kt}$$

where k is a constant, so that if $R = e^{kt}$ and k is negative

$$\frac{dR}{dt} = kR$$

so that the radiation at time t may be described by the function :

$$R = R_0 e^{kt} \tag{6.12}$$

where k is the decay constant and R_0 is the original level of radio-activity.

Half-life

The half-life of a radio-active substance is defined to be the time taken for the level of radiation to fall by one half. Thus if t_{half} is the half-life we can use *equation* (6.12) to give :-

$$\frac{R_0 e^{kt_{half}}}{R_0} = \frac{1}{2}$$

$$\therefore e^{kt_{half}} = 1/2$$
$$kt_{half} = \ln(1/2)$$
$$t_{half} = \frac{\ln(1/2)}{k}$$
$$= -0.693/k$$

The half-life of polonium 210 is 138 days. What is its decay constant?

$$k = t_{half}/-0.693 = 138/-0.693 = -0.005$$

---

[1]Bq: 1 Bequerel = 1 disintegration per second

Linear Regression - The Method of Least Squares

Here, for completeness we derive the equations used in section (4.9.2) for fitting the best straight line through a set of data points.

Suppose we have a set of experimental values

$$(x_1, y_1), (x_2, y_2), \dots (x_i, y_i) \dots (x_n, y_n)$$

and we suspect that the two variables are linearly related $(y = a + bx)$.

Draw the line y$= a + bx$ as shown (at present we don't know the values of a and b - we just guess them!).

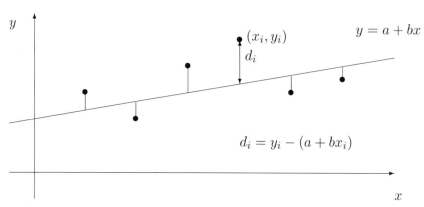

We can define the quantities d_i as the distance of the experimental points from the line (in the y direction) i.e. the difference between the experimental value of y (y_i) and the value of y predicted by the straight line $(a + bx_i)$.

$$d_i = y_i - (a + bx_i)$$

A measure of the goodness of fit can be obtained by adding the squares of all such d_i.

$$\therefore G = \sum_{i=1}^{n} d_i{}^2$$

We use the squares of the differences in order that the contribution from each difference is positive and therefore G will always be positive. If G is large the fit is bad, if G is small the fit is good.

$$
\begin{aligned}
G &= \sum_{i=1}^{n} [y_i - (a + bx_i)]^2 \\
&= \sum_{i=1}^{n} [y_i - a - bx_i]^2 \\
&= \sum_{i=1}^{n} [y_i^2 + a^2 + b^2 x_i^2 - 2y_i a - 2y_i b x_i + 2abx_i] \\
&= \sum_{i=1}^{n} y_i^2 + na^2 + b^2 \sum x_i^2 - 2a \sum y_i - 2b \sum x_i y_i + 2ab \sum x_i
\end{aligned}
$$

Note the term na^2 which is the result of adding n copies of a^2
We can minimise G in order to obtain the best fit:

   i) by changing a, keeping b constant (moving the line up or down)
   ii) by changing b, keeping a constant (rotating the line)

$$
\begin{aligned}
\frac{\partial G}{\partial a} &= 2na - 2\Sigma y + 2b\Sigma x \\
\text{at minimum } na - \Sigma y + b\Sigma x &= 0 \qquad (6.13) \\
\frac{\partial G}{\partial b} &= 2b\Sigma x^2 - 2\Sigma xy + 2a\Sigma x \\
\text{at minimum } b\Sigma x^2 - \Sigma xy + a\Sigma x &= 0 \qquad (6.14) \\
\text{from (6.13)} \quad a &= \frac{\Sigma y - b\Sigma x}{n} \qquad (6.15)
\end{aligned}
$$

substitute a from (6.15) into (6.14)

$$
\begin{aligned}
b\Sigma x^2 - \Sigma xy + \frac{(\Sigma y - b\Sigma x)\Sigma x}{n} &= 0 \\
nb\Sigma x^2 - n\Sigma xy + \Sigma y \Sigma x - b\Sigma x \Sigma x &= 0 \\
b(n\Sigma x^2 - \Sigma x \Sigma x) &= n\Sigma xy - \Sigma x \Sigma y \\
b &= \frac{n\Sigma xy - \Sigma x \Sigma y}{n\Sigma x^2 - (\Sigma x)^2} \qquad (6.16)
\end{aligned}
$$

Thus we calculate b from *equation* (6.16) and then use its value in *equation* (6.15) to calculate a.

Cylinder of Minimum Surface Area

A hot water tank consists of a closed cylinder of height H and radius R. In order to minimise heat loss, and hence save the planet and our heating bill, we would like the tank to have minimum surface area. If the volume is fixed, what ratio of H to R will give the mimimum total surface area? The same principles apply to microbial cells or cylindrical cows!

$$V = \pi R^2 H = constant \qquad (6.17)$$
$$A = 2\pi RH + 2\pi R^2 \qquad (6.18)$$
$$\text{from } equation \text{ (6.17) } H = \frac{V}{\pi R^2}$$
$$\text{Substitute in } equation \text{ (6.18) } A = 2\pi R\frac{V}{\pi R^2} + 2\pi R^2$$
$$= \frac{2V}{R} + 2\pi R^2$$
$$\frac{dA}{dR} = -\frac{2V}{R^2} + 4\pi R$$
$$\text{At minimum } \tfrac{dA}{dR} = 0 \text{, hence } 4\pi R = \frac{2V}{R^2}$$
$$\text{Replace } V \text{ } eq. \text{ (6.17) to give } 4\pi R = \frac{2\pi R^2 H}{R^2}$$
$$\text{and hence for minimum area } \frac{H}{R} = 2$$

Exercises

1. Find the following derivatives.

   (a) $\frac{d}{dt}\sin t$

   (b) $\frac{d}{dt}\cos^2 t$

   (c) $\frac{d}{dt}\sin(t^2)$

   (d) $\frac{d}{dz}3ze^z$

   (e) $\frac{d}{dx}\ln(3x^2)$

2. The milk yield of the cows in a herd has been found to follow a curve of the type

$$y = Ate^{-Bt}$$

   where $y =$ yield in litres/day and t is the time in days from the start of lactation. A specific animal was recorded as giving 15.9 litres on day 10 and 24.7 litres on day 20.

   What will be the milk yield for this animal at day 100?

   When will the maximum yield for this cow occur and what will it be?

3. (a) <u>From first principles</u> prove that $\frac{d}{dx} x^2 = 2x$

    (b) A crop is observed to be infected with a form of rust. A survey shows that the number of plants infected at present is 3 per sq. metre. The disease is known to spread according to the equation

    $$\frac{dN}{dt} = 2\sqrt{N}$$

    where N is the number of infected plants per sq. metre and t is the time in days. When the infected rate reaches $100/m^2$ it is felt that the crop cannot be saved. How many days do we have in which to effect a cure? [Hint:- the two parts of the question are related].

4. Find the derivative of each of the following

    (a) $x^{3/2}$

    (b) $x^2 e^{3x}$

    (c) $\sin 3x$

    (d) $\ln 2x$

    (e) $\tan x (= \frac{\sin x}{\cos x})$

    (f) $e^{x \sin x}$

5. Find the following derivatives

    (a) $\frac{d}{dx}(a + bx + cx^2)$

    (b) $\frac{d}{dt}(\sin wt)$

    (c) $\frac{d}{dx}(\frac{\sin x}{x})$

    (d) $\frac{d}{d\sin x}(\sin^2 x)$

    (e) $\frac{d}{dx}(3x^2 + x)\sin(2x)$

    (f) $\frac{d}{dx} e^{\cos(\ln(x))}$

6. The specific weight of water at $t°C$ is given by

    $$w = 1 + (5.3 \times 10^{-5})t - (6.53 \times 10^{-6})t^2 + (1.4 \times 10^{-8})t^3$$

    Find the temperature at which water has maximum specific weight.

7. An animal is fed on a diet in which the concentration (F) of an added factor is varied. It is found that the daily consumption (D) of the diet is related to the concentration of added factor as follows:-

    $$D = A - aF$$

    where A is the consumption when no factor is added and a is a constant. What value of F should be chosen so that the animal consumes the maximum amount of the factor?

Answers

1. (a) $\cos t$

   (b) $-2\sin t\cos t$

   (c) $2t\cos t^2$

   (d) $3ze^z + 3e^z = 3(z+1)e^z$

   (e) $2/x$

2. $A = 2.04704, B = 0.02527, 16.36$ litres, 29.8 litres on day 39

3. (a)

   (b) 8.27 days

4. (a) $\frac{3}{2}x^{1/2}$

   (b) $e^{3x}(2x + 3x^2)$

   (c) $3\cos 3x$

   (d) $\frac{1}{x}$

   (e) $\frac{1}{\cos^2 x}$

   (f) $(sinx + x\cos x)e^{x\sin x}$

5. (a) $b + 2cx$

   (b) $w\cos wt$

   (c) $\frac{\cos x}{x} - \frac{\sin x}{x^2}$

   (d) $2\sin x$

   (e) $2(3x^2 + x)\cos(2x) + (6x + 1)\sin(2x)$

   (f) $-\frac{1}{x}\sin(\ln(x))e^{\cos(\ln(x))}$

6. 4.113^oC

7. $\frac{A}{2a}$

Chapter 7

Integral Calculus

7.1 Introduction

Integration (anti-differentiation) is the inverse of differentiation. We begin with an expression defining the rate of change (growth rate, heat flux, rate of infection, glucose in and glucose out ...) and use it to define the state of the system (size, temperature, magnitude of the epidemic, amount of glucose in pool ...).

If we have two functions $y(x)$ and $g(x)$ so that

$$\frac{d}{dx} y(x) = g(x)$$

i.e. the derivative of $y(x)$ is $g(x)$, then $y(x)$ is the *integral* of $g(x)$, which we write as follows:

$$\text{if } \frac{d}{dx} y(x) = g(x) \text{ then } \int g \, dx = y + c \qquad (7.1)$$

Here c is an arbitrary constant, which must be included because the derivative of a constant is zero. *Equation* (7.1) is the basis of integral calculus. An integral, such as this, in which there is an unknown constant is known as an *indefinite integral*. The value of c can usually be found because there will be a boundary condition at which the values of both x and y will be known so that substituting these values will allow the calculation of c. Here are some examples which you have seen in the previous chapter.

$$\frac{d}{dx} x^2 = 2x \quad \therefore \quad \int 2x \, dx = x^2 + c$$

and

$$\frac{d}{dx} e^x = e^x \quad \therefore \quad \int e^x \, dx = e^x + c$$

and

$$\frac{d}{dx} \sin x = \cos x \quad \therefore \quad \int \cos x \, dx = \sin x + c$$

where c is an unknown constant.

If, at this point you say to yourself "So what?", I for one have sympathy with you - all we have gained so far is another definition, but we don't understand what it is all about. Accepting things because somebody says so is hardly science, so let us try to see why and how integration works.

First of all let us see if we can make sense of the notation:

$$\int y \, dx$$

which is read as "the integral ($\int$) of the function (y) with respect to x (dx)". Here x is an independent variable upon which the variable y depends. The above expression represents the sum of all the values of $y \, dx$ over the range of x. For example if $y = 1$, $\int 1 \, dx = \int dx = x$ which is the sum of all the infinitely small bits of x, which of course will be equal to x. Similarly $\int dy = y$ or $\int d \, anything = anything$.

7.2   Integration as the Area under a Curve

Consider a curve whose equation $y = f(x)$ is known.

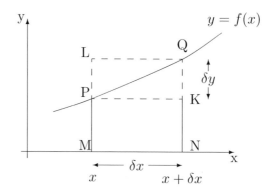

In particular, examine a small section of the curve between x and $x + \delta x$. We see that the area δA defined by the x axis, the ordinates at x and $x + \delta x$ and the curve $y = f(x)$ lies between the areas of the rectangles MLQN and MPKN:-

$$\text{Area (MPKN)} \quad < \quad \delta A \quad < \quad \text{Area (MLQN)}$$
$$y.\delta x \quad < \quad \delta A \quad < \quad (y + \delta y).\delta x$$
$$\therefore \quad y \quad < \quad \frac{\delta A}{\delta x} \quad < \quad y + \delta y$$

$$\text{Now as} \quad \delta x \to 0 \qquad \frac{\delta A}{\delta x} \to \frac{dA}{dx} = y$$

But from our definition of integration (*equation* (7.1))

$$\text{if} \quad \frac{dA}{dx} = y$$

$$A(x) = \int y\,dx \qquad (7.2)$$

i.e. the area under the curve is given by the integral of the function $y = f(x)$.

However there is still a problem with this interpretation of integration since the area calculated using *equation* (7.2) will always contain an arbitrary constant, which implies that the area can have any value!

This is not such a difficulty when we think about it more carefully, since, in order to define an area it is necessary that we define a <u>closed</u> area. It is impossible to calculate the area defined by the x-axis and $y = f(x)$ if we don't define the left and right boundaries also. Thus, when we are using integration to calculate the area under a curve we must define the boundaries, which we write as follows:

$$area = \int_a^b y\,dx \qquad (7.3)$$

where a and b are the lower and upper boundaries of the area to be found. The area under the curve is then calculated as:

$$
\begin{aligned}
area \;=\;& (A(b) + constant) - (A(a) + constant) \\
=\;& A(b) - A(a) \qquad \text{since the two values of } constant \text{ cancel}
\end{aligned}
$$

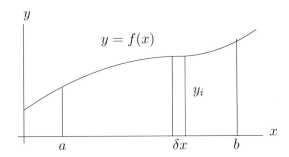

We can think of *equation* (7.3) as the sum of the areas of all strips of width δx under the curve between $x = a$ and $x = b$, when the area is made up of an infinite number of extremely narrow strips. The integral sign $\int$ is derived from a long S - a shorthand notation for "sum".

$$
\begin{aligned}
area \;=\;& \underset{n\to\infty}{Lim} \sum_{i=1}^{n} y_i.\delta x \\
& \text{where } \delta x = \frac{b-a}{n} \text{ and } y_i \text{ is the height of strip } i \\
=\;& \int_a^b y\,dx = [A(x)]_a^b \equiv A(b) - A(a) \qquad (7.4)
\end{aligned}
$$

The integral depicted above and in *equation* (7.3) is known as a *definite integral* because it does not contain an unknown constant. Take **note** of the **shorthand notation** in *equation* **(7.4)**

Example - Area under the curve $y = x$

Find the area defined by the curve $y = x$, the x axis and the ordinates $x = 1$ and $x = 3$.

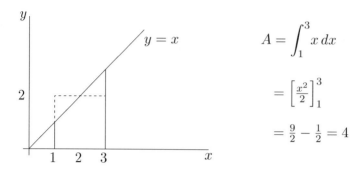

$$A = \int_1^3 x \, dx$$

$$= \left[\frac{x^2}{2} \right]_1^3$$

$$= \frac{9}{2} - \frac{1}{2} = 4$$

This is easily checked by evaluating the area directly.

Example - The area bound by $y = \cos x$ for $\frac{\pi}{2} \le x \le \pi$

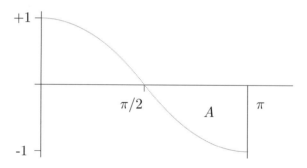

$$
\begin{aligned}
A &= \int_{\pi/2}^{\pi} \cos x \, dx \\
&= [\sin x]_{\pi/2}^{\pi} \\
&= 0 - 1 \\
&= -1
\end{aligned}
$$

Note that this "area" turns out to be negative! Treating the integral as an area is strictly incorrect, since the integral of a negative function gives a negative area. If we really need to calculate the area we must integrate the negative and positive regions of the function separately and deal with the negative integrals appropriately.

Example - $\int_0^\infty e^{-x}\,dx$

Find the area under the curve e^{-x} between $x = 0$ and $x = \infty$.

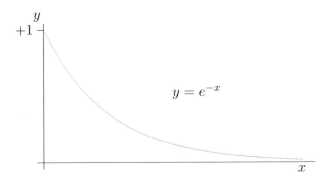

$$
\begin{aligned}
\int_0^\infty e^{-x}\,dx &= \left[-e^{-x}\right]_0^\infty \\
&= (-e^{-\infty}) - (-e^0) \\
&= 1
\end{aligned}
$$

n.b. this area would be very difficult to calculate by direct measurement!

7.2.1 Area of a Circle 1

The concept of summing fundamental elements of area can be extended further. Consider the diagram below which shows a small segment of a circle of radius r, bounded by radii at θ radians and $\theta + \delta\theta$ radians.

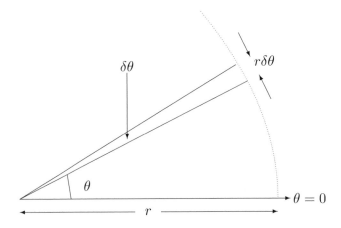

The area of such a segment provided that $\delta\theta$ is small, and we are going to let it be infinitesimally small, can be approximated by

$$
\text{area of segment} = \frac{r \times r\delta\theta}{2} = \frac{r^2\delta\theta}{2}
$$

since the small segment is approximately a triangle with sides r and $r\delta\theta$. The angles at the circumference will be approximately right-angles provided $\delta\theta$ is small.

Now, we can calculate the area of the circle by integrating this expression (summing all the area segments within the circle) as follows:

$$\text{Area of circle} = \int_0^{2\pi} \frac{r^2 \, d\theta}{2}$$

and because $r^2/2$ is constant we can move it outside the integral sign to give:

$$
\begin{aligned}
\text{Area of circle} \ &= \ \frac{r^2}{2} \int_0^{2\pi} d\theta \\
&= \ \frac{r^2}{2} \, [\theta]_0^{2\pi} \\
&= \ \frac{r^2}{2} (2\pi - 0) \\
&= \ \pi r^2
\end{aligned}
$$

7.2.2   Area of a Circle 2

In a similar way we can think of a circle of radius r to be made up of many rings defined by circles of radius x and $x + \delta x$. The area of such a ring can be approximated by $2\pi x \delta x$. If we now integrate the areas of all such rings letting x take all the values between 0 and r and $\delta x \to 0$ we have:

$$
\begin{aligned}
\text{Area of circle} \ &= \ \int_0^r 2\pi x \, dx \\
&= \ 2\pi \int_0^r x \, dx \\
&= \ 2\pi \left[\frac{x^2}{2} \right]_0^r \\
&= \ 2\pi (r^2/2 - 0) \\
&= \ \pi r^2
\end{aligned}
$$

Once again we see that there is more than one way to generate a solution. It takes a little imagination (cunning!) but when we reach the same solution via two different methods it gives us confidence in the result.

7.3 Techniques of Integration

7.3.1 The Chain Rule

The chain rule for the derivative of a function of a function is given by

$$\frac{dy}{dx} = \frac{dy}{du}\frac{du}{dx}$$
$$where \ \ y = f(u) \ and \ u = g(x)$$

The corresponding formula for integration is given by

$$\int y \, dx = \int y \frac{dx}{du} \, du$$

Here we treat the function y as a function $y = f(u)$ where $u = g(x)$, i.e. we make a substitution which hopefully will simplify the integral.

Example - $\int \sin^2 x \cos x \, dx$

$$\text{If we put} \ \ u = \sin x$$
$$\frac{du}{dx} = \cos x \quad \therefore \quad \frac{dx}{du} = \frac{1}{\frac{du}{dx}} = \frac{1}{\cos x}$$
$$\text{then} \quad \int \sin^2 x \cos x \, dx = \int u^2 \cos x \frac{1}{\cos x} \, du = \int u^2 \, du$$
$$= \frac{u^3}{3} + c$$
$$= \frac{1}{3} \sin^3 x + c$$

As with using the chain rule for differentiation I sometimes find it easier to omit the substitution and proceed as follows:

$$\int \sin^2 x \cos x \, dx$$
$$= \int \sin^2 x \, d\sin x \quad \text{because } d\sin x/dx = \cos x \text{ and hence } \cos x \, dx = d\sin x$$
$$= \frac{1}{3} \sin^3 x + c$$

Example - $\int_0^2 e^{-x^2} x \, dx$

In order to simplify the equation make the substitution -

$$u = -x^2 \quad \text{so that} \quad \frac{du}{dx} = -2x \quad \text{and } dx = \frac{du}{-2x}$$

$$\text{hence } I = \int_0^{-4} e^u x \frac{1}{-2x} \, du$$

n.b. The limits of the integration have changed since we are now integrating over the scale of u, i.e. from 0 to $-(2^2)$.

$$
\begin{aligned}
I &= -\frac{1}{2} \int_0^{-4} e^u \, du \\[2mm]
&= -\frac{1}{2} e^u \Big|_0^{-4} \\[2mm]
&= -\frac{1}{2} e^{-4} + \frac{1}{2} e^0 \\[2mm]
&= \frac{1}{2}(1 - e^{-4})
\end{aligned}
$$

Again, it sometimes simplifies the solution if we leave out the overt substitution step:

$$
\begin{aligned}
I &= \int_0^2 e^{-x^2} x \, dx \\[2mm]
&= \int_0^4 e^{-x^2} \frac{dx^2}{2} = \frac{1}{2} \int_0^4 e^{-x^2} dx^2 \\[2mm]
&= \frac{1}{2} \left[-e^{-x^2} \right]_0^4 = \frac{1}{2} \left[e^0 - e^{-4} \right] \\[2mm]
&= \frac{1}{2} \left[1 - e^{-4} \right]
\end{aligned}
$$

7.3.2   Integration by Parts

The differential of a product is defined in *equation* (6.7)

$$\frac{d}{dx} u.v = u \frac{dv}{dx} + v \frac{du}{dx} \tag{7.5}$$

$$
\begin{aligned}
e.g. \; \frac{d}{dx} x \sin x &= x \frac{d}{dx} \sin x + \sin x \frac{d}{dx} x \\[2mm]
&= x \cos x + \sin x
\end{aligned}
$$

We can integrate equation (7.5)

$$\int \frac{d}{dx} (u.v) \, dx = \int u \frac{dv}{dx} \, dx + \int v \frac{du}{dx} \, dx$$

to give

$$uv = \int u \, dv + \int v \, du$$

$$\text{or } \int u \, dv = uv - \int v \, du \tag{7.6}$$

The splitting of an integral suggested by the above formula will sometimes help to reduce the complexity of difficult integrals. It requires practice to see that this method will help when a substitution will not. There is a measure of artistry in the solution of integrals!

Example - $\int x e^x \, dx$

$$
\begin{aligned}
\text{Put } u = x \qquad dv &= e^x \, dx \\
\frac{du}{dx} = 1 \qquad \frac{dv}{dx} &= e^x \\
du = dx \qquad v &= e^x \\
\int x \, e^x \, dx &= x \, e^x - \int e^x \, dx \\
&= x e^x - e^x + c \\
&= e^x (x - 1) + c
\end{aligned}
$$

Example - $\int x^2 \sin x \, dx$

$$
\begin{aligned}
\text{Put } u &= x^2 \\
\text{so that} \frac{du}{dx} &= 2x \\
\text{and hence } du &= 2x \, dx \\[2mm]
\text{and put } dv &= \sin x \, dx \\
\text{so that } \frac{dv}{dx} &= \sin x \\
\text{and hence } v &= -\cos x
\end{aligned}
$$

Now substituting the above in equation 7.6 we have:

$$
\int x^2 \sin x \, dx = -x^2 \cos x + 2 \int x \cos x \, dx
$$

Note - this integral involves only x^1 as opposed to x^2, hence things are improving and we should proceed along the same track.

Now put $u = x$ so that $du = dx$

and $dv = \cos x \, dx$ so that $v = \sin x$

Substituting the above in equation 7.6 gives:

$$\int x\cos x\,dx = x\sin x - \int \sin x\,dx$$

$$= x\sin x + \cos x + c$$

$$\therefore \int x^2 \sin x\,dx = -x^2\cos x + 2x\sin x + 2\cos x + c'$$

$$= \cos x(2 - x^2) + 2x\sin x + c'$$

Example - The factorial function

The factorial function

$$\int_0^\infty x^n e^{-x}\,dx$$

has important applications in statistics. You will come across it when using the Poisson and Binomial distributions. You have already seen it in the context of the exponential function. When n is an integer "factorial n" can be expressed as follows:

$$n! = n \times (n-1) \times (n-2) \cdots \times 2 \times 1$$

and its value represents the number of different ways that n distinct objects can be arranged in order.

$$\text{Show that} \int_0^\infty x^n e^{-x}\,dx = n!$$

$$\int_0^\infty x^n e^{-x}\,dx = [-x^n e^{-x}]_0^\infty + \int_0^\infty nx^{n-1}e^{-x}\,dx$$

$$= 0 + n\int_0^\infty x^{n-1}e^{-x}\,dx$$

$$= n(n-1)\int x^{n-2}e^{-x}\,dx$$

$$\vdots$$

$$= n(n-1)\cdots 1\int xe^{-x}\,dx$$

$$\int_0^\infty xe^{-x}\,dx = -xe^{-x}]_0^\infty + \int_0^\infty e^{-x}\,dx$$

$$= 0 - [e^{-x}]_0^\infty$$

$$= -e^{-\infty} + e^0 = 1$$

$$\therefore \int_0^\infty x^n e^{-x}\,dx = n!$$

7.4 Summary Notes on Integration

7.4.1 Standard Integrals

$$\int x^n \, dx = nx^{n+1} + c \qquad \text{provided } n \neq -1$$

$$\int x^{-1} \, dx = \ln x + c$$

$$\int \sin x \, dx = -\cos x + c$$

$$\int \cos x \, dx = \sin x + c$$

$$\int e^x \, dx = e^x + c$$

where c is an unknown constant.

7.4.2 Techniques

If $\int f(x) \, dx = g(x)$ then

$$\int f(ax) \, dx = \frac{1}{a} g(x)$$

If $y = f(u)$ and $u = g(x)$ then

$$\int y \, dx = \int y \frac{dx}{du} \, du$$

If both u and v are functions of x then

$$\int u \, dv = uv - \int v \, du$$

7.5   Applications

Mean Value

One very useful result of integration is:-

$$\frac{\int_a^b f(x)\,dx}{b-a} = \text{mean value of } f(x)$$

Example - The mean value of $y = x$ for $a \leq x \leq b$.

If $y = f(x) = x$ then the mean value of y over the interval $a \leq x \leq b$ is:

$$\bar{y} = \int_a^b \frac{x\,dx}{b-a} = \frac{\left[\frac{x^2}{2}\right]_a^b}{b-a}$$

$$= \frac{b^2 - a^2}{2(b-a)} = \frac{(b-a)(b+a)}{2(b-a)} = \frac{b+a}{2}$$

Surfaces and Volumes of Revolution

We have seen that the integral can be interpreted as the sum of an infinite number of strips in determining the area under the curve.

We can extend this idea to find the volume, surface area etc. of surfaces of revolution. A surface of revolution is created by rotating a two-dimensional object about an axis.

As an illustration let us calculate the volume of a circular cone. The cone could be generated by "revolving" a triangle about the x - axis. The cone has a base radius R, on the right, and its height will be H, though I have drawn it on its side, and its apex is at the origin.

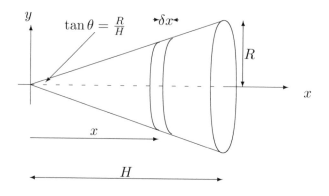

Consider a slice of thickness δx perpendicular to the x-axis at a distance x from the origin. This slice will be a disc of radius $x \tan \theta$ and thickness δx. Its volume will be approximately

$$\pi (x \tan \theta)^2 \delta x$$

Here we have ignored the fact that the edge of the disc is sloping. The error involved is of the order δx^2 (the actual term is $2\pi\theta^2\delta x^2$) and is negligible in comparison to the volume of the disc.

Now if we add together the volumes of all such discs in the range $0 < x < H$ we can calculate the volume of the cone.

$$
\begin{aligned}
V &= \int_0^H \pi (x \tan \theta)^2 \, dx = \pi \tan^2 \theta \int_0^H x^2 \, dx \\
&= \pi \tan^2 \theta \left[\frac{x^3}{3} \right]_0^H \\
&= \frac{1}{3}\pi \tan^2 \theta H^3 \\
\text{but } H \tan \theta &= R \\
\therefore \ V &= \frac{1}{3}\pi R^2 H
\end{aligned}
$$

Equations of Motion

The following are equations governing the motion of a particle undergoing uniform acceleration:-

$$
\begin{aligned}
s &= ut + \frac{1}{2}ft^2 \\
v^2 &= u^2 + 2fs \\
v &= u + ft \\
\textit{where} \quad s &= \text{distance travelled at time } t \\
v &= \text{velocity} \\
f &= \text{acceleration } (= constant) \\
u &= \text{initial velocity}
\end{aligned}
$$

The above equations can be derived directly from Newton's second law of dynamics. This law states that "the rate of change of momentum is proportional to the applied force, and is in the direction in which the force acts" and can be written mathematically as follows:

$$\frac{d}{dt}(mv) = P$$

where $m = $ mass, $v = $ velocity and $P = $ force, in appropriate units.

$$\frac{d}{dt}(mv) = m\frac{dv}{dt} + v\frac{dm}{dt} = P$$

In general it may be assumed that m is constant and that the equation therefore reduces to

$$\frac{d}{dt}(mv) = m\frac{dv}{dt}$$

The assumption (*that mass is invariant*) held back the development of physics for a long time until Poincaré and Einstein questioned it to produce the theory of relativity.

If we express the force as $P = mf$ where f is a constant with appropriate units we now have

$$m\frac{dv}{dt} = mf$$

$$\therefore \quad \frac{dv}{dt} = f \qquad \text{i.e. acceleration} = \text{constant}$$

$$\text{integrating gives} \quad v = ft + c$$

but if $v = u$ when $t = 0$ it follows that $c = u$

$$\therefore \quad v = u + ft = \frac{ds}{dt} \qquad (7.7)$$

$$\text{and } s = \int (u + ft)dt$$

$$= ut + \frac{1}{2}ft^2 + d$$

$$\text{if } s = 0 \text{ when } t = 0 \text{ then } d = 0$$

$$\therefore \quad s = ut + \frac{1}{2}ft^2 \qquad (7.8)$$

$$\text{Also} \quad v = u + ft$$

$$\therefore \quad v^2 = u^2 + 2ftu + f^2t^2$$

$$= u^2 + 2f(ut + \frac{ft^2}{2})$$

$$\text{substituting (7.8) gives}$$

$$v^2 = u^2 + 2fs \qquad (7.9)$$

Pollution of a Lake

A lake has volume $V(m^3)$ and is fed by a river which flows through at a rate $R(m^3/sec)$. If a small volume $P_0(m^3)$ of pollutant is accidentally dropped into the lake, derive an expression for the amount of pollutant in the lake at subsequent time t. Assume perfect mixing occurs.

Find an expression for the time when the concentration of pollutant in the lake has fallen to one tenth of its initial value.

Let the amount of pollutant at time t be $P(t)$, so that the rate of loss of pollutant from the lake at any time t will be the concentration $(P(t)/V)$ multiplied by the outflow rate:-

$$\frac{dP}{dt} = -\frac{P(t)}{V} \times R = -\frac{R}{V} \times P(t) \tag{7.10}$$

Observation that dP/dt is proportional to P should sound *exponential* warning bells. Equations of the form:

$$\frac{dy}{dx} = ky \tag{7.11}$$

are satisfied by functions of the type $y = e^{kx}$, or more generally by:

$$y = Ae^{kx} \tag{7.12}$$

where A and k are constants.

Comparison of *equation* (7.10) with *equation* (7.11) and *equation* (7.12) leads by analogy to the following solution:

$$P(t) = Ae^{-(R/V)t} \tag{7.13}$$

However when $t = 0$ we know that $P(0) = P_0$, hence $A = P_0$ and we have:

$$P(t) = P_0e^{-(R/V)t} \tag{7.14}$$

The term (R/V) is often referred to as the *rate constant* in this sort of problem.

The second part of the problem asks when the amount is reduced by a factor of 10. Thus, we require t such that

$$\frac{P(t)}{P_0} = 0.1$$

Hence

$$\frac{P_0e^{-(R/V)t}}{P_0} = 0.1$$
$$e^{-(R/V)t} = 0.1$$
$$-(R/V)t = \ln(0.1)$$
$$t = -(V/R)\ln(0.1)$$

Exercises

1. Calculate each of the following indefinite integrals.

   (a) $\int x^2 \, dx$

   (b) $\int \cos x \, dx$

   (c) $\int (a + bx + cx^2) \, dx$

   (d) $\int \frac{1}{x} \, dx$

   (e) $\int x^2 \, dx^2$

2. A spanner was dropped from the top of a building.

   (a) Assuming that the gravitational constant g is $10ms^{-1}$ its velocity v at time t will be given by $v = 10t$ metres per second. If it took 3 seconds to reach the ground, how high is the building?

   (b) If the spanner has mass m its potential energy at the top of the building is given by mgh and its kinetic energy just before hitting the ground at velocity v by $mv^2/2$. Evaluate both values and comment on your results.

3. The surface area A of a sphere of radius r is given by

$$A = 4\pi r^2$$

   where r is the radius.

   (a) If a spherical cell of radius r increases its radius by δr, what will be its increase in volume?

   (b) using the above result find an expression for the volume of the cell if its radius is R.

4. A radioactive source decays according to the relationship

$$C = C_0 \, e^{-kt}$$

   where C is the rate of emission at time t. C_0 is the original source strength and k is a constant. Derive an expression for R if

$$R = \int_0^T C_0 \, e^{-kt} \, dt$$

   Sketch a graph of R against T and explain the result. In particular, what is the significance of the expression C_0/k?

Answers

1. (a) $\frac{x^3}{3} + c$

   (b) $\sin x + c$

   (c) $ax + \frac{bx^2}{2} + \frac{cx^3}{3} + d$

   (d) $\ln x + c$

   (e) $\frac{x^4}{2} + c$ hint: $\frac{(x^2)^2}{2} + c$

2. (a) 45 metres

   (b) both values are $450m$. The potential energy lost in falling is converted to kinetic energy.

3. (a) $4\pi r^2 \delta r$

   (b) $4\pi r^3 / 3$

4. The integral is

$$R = \frac{C_0}{k}(1 - e^{-kT})$$

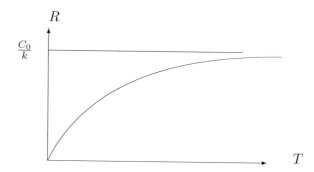

$R(T)$ represents the total radiation emitted during the period $0 - T$.

The expression C_0/k is the total radiation to be emitted. Note that the larger k, the faster the decay, and the smaller the total radiation.

Chapter 8

Matrix Algebra

8.1 Introduction

Why matrix algebra? It looks complicated, and indeed some of it can be difficult - but a little knowledge goes a long way. Could you solve 50 simultaneous equations with 50 unknowns (nutritional properties of a diet related to ingredients for example)? You can if you want to, almost without effort - provided you have access to a computer. It's a case of some of the underlying maths being difficult, but making use of it is easy - once you know how to go about it.

Many modern scientific analyses could not be achieved without the use of matrices, therefore it will be of use to gain a little familiarity with some of the terms. Don't be afraid, it's easier than it looks!

The objectives of this chapter are to give you sufficient knowledge to get the results, rather than to show you how it works. But it is also enlightening to see how some mathematics results from a need, rather than from a purely academic exercise.

You will be familiar with problems like the following in which we have two equations in two unknowns.

$$3x + 4y = 18 \qquad (8.1)$$
$$2x - y = 1 \qquad (8.2)$$

You may also remember that in order to find a solution (the values of x and y which satisfy the equations) you need to perform some fairly tedious arithmetic. This involves reducing the equations to a single equation in one unknown and then back-substituting to find the value of the other unknown.

The above is a simple example, which can be solved fairly easily. Add $4\times$ (8.2) to (8.1) to give

$$11x = 22$$

and hence $x = 2$. Substituting this value back into *equation* (8.2) produces the remaining value $y = 3$.

However when there are more equations the problem of writing down and solving such problems becomes much more difficult.

Mathematicians used to write sets of simultaneous equations as follows:

$$a_{11}x_1 + a_{12}x_2 + a_{13}x_3 + \cdots + a_{1n}x_n = b_1 \qquad (8.3)$$
$$a_{21}x_1 + a_{22}x_2 + a_{23}x_3 + \cdots + a_{2n}x_n = b_2$$
$$\vdots \qquad = \qquad \vdots$$
$$a_{n1}x_1 + a_{n2}x_2 + a_{n3}x_3 + \cdots + a_{nn}x_n = b_n$$

where the a_{ij} are coefficients of a set of unknowns x_i and the b_i are values for the right hand sides of the set of equations. Working with this notation can be very tedious, very boring, and extremely prone to errors. And so mathematicians, developed a shorthand, *matrix algebra*, whereby the above equations became:

$$AX = B \qquad (8.4)$$

in which A, X and B represent complex objects.

$$A = \begin{bmatrix} a_{11} & a_{12} & \cdots & a_{1n} \\ a_{21} & a_{22} & \cdots & a_{2n} \\ \vdots & \vdots & & \vdots \\ a_{n1} & a_{n2} & \cdots & a_{nn} \end{bmatrix}, \quad X = \begin{bmatrix} x_1 \\ x_2 \\ \vdots \\ x_n \end{bmatrix} \quad \text{and} \quad B = \begin{bmatrix} b_1 \\ b_2 \\ \vdots \\ b_n \end{bmatrix}$$

A is a square matrix with n rows and n columns. X and B are single column matrices (called column vectors) , each with n elements.

Equations 8.1 and 8.2 could also be represented by *equation* (8.4) with

$$A = \begin{bmatrix} 3 & 4 \\ 2 & -1 \end{bmatrix}, \quad X = \begin{bmatrix} x \\ y \end{bmatrix} \quad \text{and} \quad B = \begin{bmatrix} 18 \\ 1 \end{bmatrix}$$

Notice that the equivalence of equations (8.3) and (8.4), and of equations (8.1, 8.2) and (8.1) define the process of matrix multiplication. See section (8.2.7).

8.2 What is a Matrix

A matrix is a rectangular array of numbers represented as follows:

$$
A = \begin{bmatrix}
a_{11} & a_{12} & \cdots & a_{1n} \\
a_{21} & a_{22} & \cdots & a_{2n} \\
\vdots & \vdots & & \vdots \\
a_{m1} & a_{m2} & \cdots & a_{mn}
\end{bmatrix}
$$

Matrices are normally represented by capital letters. The individual numbers a_{ij} are called elementsand when a matrix is written out in full, the elements are enclosed within square brackets. The subscripts i and j identify the row and column in which the element is located. Thus a_{23} is the element which is located in the second row and the third column of A.

A row of a matrix contains all the elements whose first subscript is the same. A column of a matrix contains all the elements whose second subscript is the same.

A matrix which has m rows and n columns is called a matrix of order (m, n) or an $m \times n$ ("m by n") matrix. If $m = n$ then the matrix is said to be square and may be referred to as a matrix of order n or an n-square matrix.

It is sometimes convenient to abbreviate the matrix representation of the above to

$$
A = [a_{ij}]
$$

In the following notes we shall develop an algebra which allows us to manipulate matrices in much the same way that we can perform arithmetic on ordinary numbers (scalars). You should note that there is nothing mysterious about this, we are simply developing a notation which will allow us to write down the steps in calculations which involve groups of similar items or equations.

8.2.1 Equality oɪ ᴍatrices

Two matrices $A(m, n)$, $B(m, n)$ are said to be equal if $a_{ij} = b_{ij}$ for all i, j. This means that A and B are only equal if <u>all</u> equivalent elements are equal.

8.2.2   Addition of Matrices

If $P = [p_{ij}]$ and $Q = [q_{ij}]$ are matrices of order $m \times n$ we define the sum of P and Q to be

$$
\begin{aligned}
P + Q &= [p_{ij} + q_{ij}] \\
&= \begin{bmatrix}
p_{11} + q_{11} & p_{12} + q_{12} & \cdots & p_{1n} + q_{1n} \\
p_{21} + q_{21} & p_{22} + q_{22} & \cdots & p_{2n} + q_{2n} \\
\vdots & & & \\
p_{m1} + q_{m1} & p_{m2} + q_{m2} & \cdots & p_{mn} + q_{mn}
\end{bmatrix}
\end{aligned}
$$

Notes

$P + Q = Q + P$ the commutative law of addition.

$P + (Q + R) = (P + Q) + R$ the associative law of addition.

8.2.3   Subtraction of Matrices

$$P - Q = [p_{ij} - q_{ij}]$$

Note: we can only add and subtract matrices of the same order.

8.2.4   Zero or Null Matrix

A matrix all of whose elements are zero is called the zero or null matrix and is usually denoted by O.

$$A + O = A$$

The negative of a matrix A is defined to be $-A = [-a_{ij}]$ i.e. the negative of A is formed by changing the sign of every element of A. Thus

$$A + (-A) = O$$

8.2.5   Identity Matrix

The identity matrix is a square matrix with all its diagonal elements equal to unity and all its other elements equal to zero.

$$I = [i_{pq}] \text{ where } i_{pq} = \begin{cases} 1 & \text{if } p = q \\ 0 & \text{if } p \neq q \end{cases}$$

e.g. the identity matrix of order 3 is:

$$I = \begin{bmatrix} 1 & 0 & 0 \\ 0 & 1 & 0 \\ 0 & 0 & 1 \end{bmatrix}$$

The identity matrix has the property

$$IX = X$$

Its behaviour is similar to the number 1 in ordinary algebra.

8.2.6 Multiplication by a scalar

If k is a scalar (a number) we define $kA \equiv [ka_{ij}]$.

Thus

$$2 \begin{bmatrix} a & b \\ c & d \end{bmatrix} = \begin{bmatrix} 2a & 2b \\ 2c & 2d \end{bmatrix} = \begin{bmatrix} a & b \\ c & d \end{bmatrix} + \begin{bmatrix} a & b \\ c & d \end{bmatrix}$$

Note that every element is multiplied by the value of the scalar.

8.2.7 Multiplication of Matrices

If

$$A = \begin{bmatrix} 6 & 2 & 1 \\ 3 & 1 & 2 \\ 2 & 4 & 1 \end{bmatrix} \text{ and } B = \begin{bmatrix} 3 & 6 \\ 2 & 1 \\ 1 & 2 \end{bmatrix}$$

and $C = AB$, we define c_{ij} by multiplying the elements in the ith row of A (left to right) by the corresponding elements in the jth column of B (top to bottom) and summing the products.

$$\text{Thus } c_{11} = \begin{bmatrix} 6 & 2 & 1 \end{bmatrix} \begin{bmatrix} 3 \\ 2 \\ 1 \end{bmatrix} = 6 \times 3 + 2 \times 2 + 1 \times 1 = 23$$

$$\text{and } c_{21} = 3 \times 3 + 1 \times 2 + 2 \times 1 = 13$$

$$\text{Therefore } C = AB = \begin{bmatrix} 23 & 40 \\ 13 & 23 \\ 15 & 18 \end{bmatrix}$$

Note: AB only exists if the number of columns in A equals the number of rows in B.

In the example above we say that A is post-multiplied by B and that B is pre-multiplied by A.

If AB exists then A is said to be conformable to B for multiplication. The fact that A is conformable to B for multiplication does not mean that B is conformable to A for multiplication, as can be seen from the previous example.

Also, if A and B are square matrices of the same order, AB is not necessarily equal to BA, so that matrices do not obey the commutative law of multiplication.

8.2.8   Using matrix multiplication to rotate coordinates

Consider the point (1,0) in Cartesian co-ordinates represented as a column vector

$$\begin{bmatrix} 1 \\ 0 \end{bmatrix}$$

Pre-multiplying by the rotation matrix as follows:

$$\begin{bmatrix} \cos\theta & -\sin\theta \\ \sin\theta & \cos\theta \end{bmatrix} \begin{bmatrix} 1 \\ 0 \end{bmatrix} = \begin{bmatrix} \cos\theta \\ \sin\theta \end{bmatrix}$$

has the effect of rotating the point (1,0) anti-clockwise about the origin by an angle θ so that its new position is $(\cos\theta, \sin\theta)$.

If we now rotate the point again, this time by an angle ϕ

$$\begin{bmatrix} \cos\phi & -\sin\phi \\ \sin\phi & \cos\phi \end{bmatrix} \begin{bmatrix} \cos\theta \\ \sin\theta \end{bmatrix} = \begin{bmatrix} \cos\phi\cos\theta - \sin\phi\sin\theta \\ \sin\phi\cos\theta + \cos\phi\sin\theta \end{bmatrix}$$

The two rotations have the same effect as rotating by an angle $\theta + \phi$, so that the new coordinates will be:

$$\begin{bmatrix} \cos(\theta + \phi) \\ \sin(\theta + \phi) \end{bmatrix} = \begin{bmatrix} \cos\phi\cos\theta - \sin\phi\sin\theta \\ \sin\phi\cos\theta + \cos\phi\sin\theta \end{bmatrix} \tag{8.5}$$

The above relationships are difficult to prove without recourse to matrix methods, but are extremely useful.

8.2.9   Transpose

The transpose of a matrix A is obtained by interchanging the rows and columns and is denoted by A^T.

e.g.

$$\text{if } A = \begin{bmatrix} 1 \\ 2 \\ 3 \end{bmatrix} \text{ then } A^T = \begin{bmatrix} 1 & 2 & 3 \end{bmatrix}$$

$$\text{if } B = \begin{bmatrix} 1 & 2 \\ 3 & 4 \end{bmatrix} \text{ then } B^T = \begin{bmatrix} 1 & 3 \\ 2 & 4 \end{bmatrix}$$

8.3   Determinants

The *determinant* of a square matrix is a specific scalar, i.e. a single number, which is denoted by $\det(A)$ or $|A|$ and is closely associated with the matrix inverse. If $|A|$ is non-zero then A will have an inverse. If $|A| = 0$ the matrix does not have an inverse and is said to be singular

The determinant of a 2×2 matrix is calculated as follows:

$$\begin{vmatrix} a & b \\ c & d \end{vmatrix} = ad - bc$$

8.3.1 The determinant of a 3×3 matrix

$$\begin{vmatrix} a & b & c \\ d & e & f \\ g & h & i \end{vmatrix} = a \begin{vmatrix} e & f \\ h & i \end{vmatrix} - b \begin{vmatrix} d & f \\ g & i \end{vmatrix} + c \begin{vmatrix} d & e \\ g & h \end{vmatrix}$$

$$= aei - afh - bdi + bfg + cdh - ceg$$

8.3.2 Minors and Cofactors

Let A be an n-square matrix.

If M_{ij} is the $(n-1)$-square matrix obtained by deleting the ith row and j^{th} column from A the determinant of M is called the minor of element a_{ij}. The cofactorof a_{ij}, denoted by A_{ij} is given by

$$A_{ij} = (-1)^{i+j}|M_{ij}|$$

Note thatA_{ij} is a scalar.

We can form a matrix of cofactors from any square matrix, e.g.

$$\text{If} \quad A = \begin{bmatrix} 1 & 2 & 3 \\ 4 & 1 & 0 \\ 2 & 0 & 1 \end{bmatrix} \quad \text{the matrix of cofactors is} \quad \begin{bmatrix} 1 & -4 & -2 \\ -2 & -5 & 4 \\ -3 & 12 & -7 \end{bmatrix}$$

The determinant can be evaluated using the above notation by taking any row or column of the matrix and summing the products of each element and its cofactor.

$$|A| = \sum_{j=1}^{n} a_{ij} A_{ij} \qquad \text{for row } i$$

$$\text{or}$$

$$= \sum_{i=1}^{n} a_{ij} A_{ij} \qquad \text{for column } j$$

The determinant of matrix A above is -13.

More generally, if B is of order 3

$$|B| = -b_{12} \begin{vmatrix} b_{21} & b_{23} \\ b_{31} & b_{33} \end{vmatrix} + b_{22} \begin{vmatrix} b_{11} & b_{13} \\ b_{31} & b_{33} \end{vmatrix} - b_{32} \begin{vmatrix} b_{11} & b_{13} \\ b_{21} & b_{23} \end{vmatrix}$$

by expansion based on the second column.

Note. It pays to choose a row or column with many zeros.

8.3.3   Area of a Triangle

The area of a triangle whose vertices represented in Cartesian co-ordinates are $(x_1, y_1), (x_2, y_2)$ and (x_3, y_3) is given by:

$$\text{Area} = \frac{1}{2} \begin{vmatrix} 1 & 1 & 1 \\ x_1 & x_2 & x_3 \\ y_1 & y_2 & y_3 \end{vmatrix}$$

e.g. The area of the triangle with co-ordinates (0,0),(1,0),(0,1) will be:

$$\text{Area} = \frac{1}{2} \begin{vmatrix} 1 & 1 & 1 \\ 0 & 1 & 0 \\ 0 & 0 & 1 \end{vmatrix} = \frac{1}{2} \times \left(1 \times \begin{vmatrix} 1 & 0 \\ 0 & 1 \end{vmatrix} - 0 \times \begin{vmatrix} 1 & 1 \\ 0 & 1 \end{vmatrix} + 0 \times \begin{vmatrix} 1 & 1 \\ 1 & 0 \end{vmatrix} \right) = \frac{1}{2}$$

8.3.4   Some properties of Determinants

1. The determinant of a matrix is equal to the sum of the products obtained by multiplying the elements of any row (or column) by their respective co-factors.

2. If two rows (or columns) of a determinant are interchanged the value of the determinant changes sign.

3. A determinant in which all the elements of a row (or column) are zero has the value zero.

4. A determinant in which all corresponding elements in any two rows (or columns) are equal has the value zero.

5. The value of a determinant is unaltered by adding to the contents of any row (column) a constant multiple of the corresponding elements of any other row (column).

6. The value of a determinant is unaltered if rows and columns are interchanged.

8.4   The Inverse Matrix

For any scalar $x \neq 0$ there exists another scalar x^{-1} which we call the reciprocal or inverse of x, such that

$$x.x^{-1} = 1$$

By analogy we might expect that, for a given matrix A, we could find another matrix, which we would call A^{-1} such that

$$A.A^{-1} = I \tag{8.6}$$

We call the matrix A^{-1} (if it exists!) the inverse matrix of A. It is possible to calculate the inverse of a matrix by hand, but in general it is impractical for anything larger than a 4×4 matrix. Virtually every computer system will have access to a procedure for carrying out the calculation.

The practical use of the matrix inverse can not be over-stressed, since this, together with modern computing facilities, gives us the power to solve problems that were hitherto impossible. There are textbooks devoted to its calculation, but it is not necessary to know how the calculation is done, simply that it <u>can</u> be done.

However, some matrices do not have an inverse. They can usually be identified because they will have a determinant whose value is zero. Such a matrix is referred to as being singular . (The analogy with x and x^{-1} continues to hold, since the inverse of the number zero does not exist either!) Singularity is usually due to the rows not being independent of each other, i.e. one of the rows can be formed as a combination of some of the other rows. In this case the offending row (equation) must be replaced if possible, by a new one that is independent.

8.4.1 Solution of Linear Simultaneous Equations

We can use the inverse matrix in order to find the solution of a set of simultaneous equations.

e.g.

$$3x + 4y = 32$$

$$2x + 3y = 23$$

may be expressed as:

$$\begin{bmatrix} 3 & 4 \\ 2 & 3 \end{bmatrix} \begin{bmatrix} x \\ y \end{bmatrix} = \begin{bmatrix} 32 \\ 23 \end{bmatrix}$$

or in matrix notation as:

$$AX = C$$

so that, if we can evaluate A^{-1}, we can proceed as follows:

$$\begin{aligned} \text{Pre-multiplying by } A^{-1} \quad A^{-1}AX &= A^{-1}C \\ IX &= A^{-1}C \\ \therefore \quad X &= A^{-1}C \end{aligned}$$

In this case the inverse matrix is

$$A^{-1} = \begin{bmatrix} 3 & -4 \\ -2 & 3 \end{bmatrix}$$

$$\therefore \begin{bmatrix} x \\ y \end{bmatrix} = \begin{bmatrix} 3 & -4 \\ -2 & 3 \end{bmatrix} \begin{bmatrix} 32 \\ 23 \end{bmatrix} = \begin{bmatrix} (3 \times 32) + (-4 \times 23) \\ (-2 \times 32) + (3 \times 23) \end{bmatrix} = \begin{bmatrix} 4 \\ 5 \end{bmatrix}$$

The solution of 100 equations with 100 unknowns is not exceptional these days. Its solution on a computer would be performed in a matter of a few seconds.

8.5 Applications

8.5.1 Application to Population Dynamics

The usual basis for the description of population growth is the exponential equation

$$N_t = N_0 \, e^{rt}$$

in which N_0 is the initial population and r is the "intrinsic rate of natural increase of the population". This model is limited, in that it is restricted to the situation where all individuals within the population are identical. If the model is extended (as it may be) to include different classes within the population the mathematics becomes very complicated and the resulting model relies heavily upon the solution of integral equations.

It is possible however to describe a population in terms of several different age groups in a rather elegant method which makes use of matrix algebra. Here, we represent the population at a given time by a vector, the elements of which are the number of individuals in each age class.

$$\begin{bmatrix} n_1 \\ n_2 \\ n_3 \end{bmatrix}$$

In order to generate the population one time interval later, we multiply this vector by a square matrix as shown:

$$\begin{bmatrix} f_1 & f_2 & f_3 \\ p_1 & 0 & 0 \\ 0 & p_2 & 0 \end{bmatrix} \begin{bmatrix} n_1 \\ n_2 \\ n_3 \end{bmatrix} = \begin{bmatrix} f_1 n_1 + f_2 n_2 + f_3 n_3 \\ p_1 n_1 \\ p_2 n_2 \end{bmatrix}$$

This matrix is known as the Leslie Matrix and its elements correspond to the fecundities and survival rates for the individual age classes. It can be seen that elements in the top row give contributions to the youngest age group - the new value of n_1. The fecundity values (f_i) represent the number of offspring born to each individual in the relevant age group. The elements (p_i) below the diagonal have the effect of moving members from one age group to the next higher age group; they correspond to the survival rates for each age class.

As an example consider the population described by the following Leslie Matrix:

$$\begin{bmatrix} 0 & 9 & 12 \\ \frac{1}{3} & 0 & 0 \\ 0 & \frac{1}{2} & 0 \end{bmatrix}$$

with an initial population containing one individual in the eldest group:

$$\begin{bmatrix} 0 \\ 0 \\ 1 \end{bmatrix}$$

Successive generations of the population are predicted by repeated multiplication as follows:-

time		0	1	2	3	4	5	6	7
	young	0	12	0	36	24	108	144	372
population	middle	0	0	4	0	12	8	36	48
	old	1	0	0	2	0	6	4	18
	total	1	12	4	38	36	122	184	438

Initially the population structure oscillates but it can be seen that the relative sizes of the age groups is stabilising and that the population increase in each generation seems to be settling down to some steady value.

The ratio of numbers within each age class gradually stabilises (in fact the ultimate ratio is 24 : 4 : 1) and it can be seen that the population doubles at each time interval. This factor corresponds to an intrinsic rate of increase of 2.

It is possible to predict the rate of increase and the stable population patterns from the Leslie Matrix without repeated multiplication, these properties are related to the eigenvalues and vectors of the matrix - see the following section.

8.5.2 Eigenvalues and eigenvectors

Eigenvalues and their corresponding vectors are closely linked to properties associated with the system described by the matrix equations. Thus in the above application the vector of population size, and the fact that the population doubles at each time step is linked to the dominant eigenvalue and its associated eigenvector.

A square matrix A is said to have an *eigenvalue* λ and corresponding *eigenvector* X if

$$AX = \lambda X \qquad (8.7)$$

A matrix of order n will have n eigenvalues, not necessarily different, and n eigenvectors. Eigenvectors are not uniquely defined since, if X is an eigenvector, so is any scalar multiple cX. Eigenvectors are usually scaled so that $\Sigma x_i^2 = 1$.

If we rearrange equation 8.7 as follows

$$(A - \lambda I)X = 0 \tag{8.8}$$

where I is the unit matrix, we have a system of homogeneous linear equations (the equations all have zero on the right hand side). A necessary condition for the non-trivial solution of such a set of equations is that the determinant of the coefficients is zero. Thus

$$\det(A - \lambda I) = \begin{vmatrix} a_{11} - \lambda & a_{12} & \cdots & a_{1n} \\ a_{21} & a_{22} - \lambda & \cdots & a_{2n} \\ \vdots & \vdots & & \vdots \\ a_{n1} & a_{n2} & \cdots & a_{nn} - \lambda \end{vmatrix} = 0 \tag{8.9}$$

The above is called the *characteristic* (or sometimes *secular*) equation of the matrix A and is a polynomial in λ which will have n roots, from which we can find the n eigenvalues.

Eigenvalues and vectors give access to a great many applications. For instance the eigenvectors of a variance-covariance matrix form the basis of principle component analysis. They also show the possible stable states of a system, as in the Leslie matrix example above.

Chapter 9

The End of the Beginning

If you have struggled to read this far - well done! I hope the struggling was not in vain and that you have found the effort rewarding.

If you have been able to understand most of the content - you are obviously reasonably competent with mathematics - as with any language, I hope that you will use what you know, otherwise you will lose it.

You have learnt sufficient to be useful and should be able to cope with most of the mathematics that bioscience will require of you. At least you can speak the language sufficiently well to be able to hold a conversation, and you have learnt a little about the way mathematicians think about problems. You will know for example that there is usually more than one method of solution, and that these methods allow you to see different aspects of the problem and its behaviour.

There is much more, but having understood the basics of a few techniques you are now in a position to look at others that are relevant to the work that you do. Do not be afraid. In only understanding a little maths you are not alone. It is generally accepted that the last person to have a good general understanding of mathematics was the Frenchman J. H. Poincaré roughly one hundred years ago.

When you come across a mathematical problem don't ignore it, try solving it - you may surprise yourself. Even if you can't find a solution, you will gain from trying, and you will gain the admiration of a mathematician if you ask for help in a way that s/he can understand.

The scientific community has hardly touched the problems that computers and numerical techniques will allow us to investigate - I hope that some of you will take up the challenge.

Good luck!

9.1   Further Reading

I wouldn't claim to have read every mathematics (or even introductory mathematics) book, but the following is a selection from those that I have found useful. I don't recommend that any student should purchase a particular book, because individual students respond in different ways to different authors. My advice is to look around and work with several, until you find one that is helpful and that you are comfortable with.

Foundation Maths by Croft A., and Davison R., Longman, 1995, ISBN0-582-23185-X. Lots of worked examples and exercises with answers.

Easy Mathematics for Biologists by Foster P. C., Harwood, 1998, ISBN 90-5702-339-3.

Catch up Maths and Stats for the Life Sciences by Harris M., Taylor G. and Taylor J., Scion, 2005, ISBN 1-904842-10-0. A useful introduction to basic mathematics directed specifically to the life and medical sciences.

Essential Mathematics and Statistics for Science by Currell G. and Dowman A., Wiley, 2005, ISBN 0-470-02229-9. A good coverage of mathematics relevant to science - especially geared to applications, with good examples based on analyses of experimental results. The latter half dealing with statistics is especially useful and clear.

Calculus made Easy by Silvanus P Thompson, Macmillan, ISBN 0 333 07445 9. It's an old book with many editions; but you can still find copies - there are four of mine somewhere! Worth reading, if only for the introduction and the epilogue. But the rest is good too and the book goes a long way to doing what the title says!

Matrix Computation for Scientists and Engineers by Alan Jennings, Wiley, 1977. ISBN 0 471 994219. Another old book, but with lots of examples and applications making use of matrix algebra. A good book to dip into for inspiration. His descriptions are clear, informative and understandable.

Fermat's Last Theorem by Simon Singh, Harper Collins, 2002, ISBN 1841157910. A good read, taking a tour through the history of mathematics to the solution by Andrew Wiles in 1995. It's fascinating and readable, you will surprise yourself and enjoy the journey.

Index